THE
FORTH
DIMENSION

A HITCHHIKER'S GUIDE
TO THE UNIVERSE

RON FORTH

 FriesenPress

One Printers Way
Altona, MB R0G 0B0
Canada

www.friesenpress.com

Editor: John van Leeuwen
Illustrator: Veronica N. Forth

The author is responsible for original ideas, and any errors
or omissions, that may or may not have crept into the text.

ISBN
978-1-03-918873-0 (Hardcover)
978-1-03-918872-3 (Paperback)
978-1-03-918874-7 (eBook)

1. SCIENCE, GRAVITY

Distributed to the trade by The Ingram Book Company

For Marla, Sabrina, Calista, and Veronica,
the four dimensions of my Universe.

TABLE OF CONTENTS

PREFACE

Every aspiring writer wants to produce a classic, a great literary novel, or a wise technical dissertation. The writer attempts to paint a crystal-clear portrait in the reader's mind, evoke subtle emotions with memorable phrases, select the perfect words to precisely transfer information from author to audience. Well, I wanted to do that too, but you will have to contend with this book instead.

This is not your father's physics book, unless you happen to be one of my three daughters. My objective is not to repeat explanations of physical phenomena that have been written about at length, in every dimension, in countless books. I want to introduce alternative perspectives on the causes of what is actually observed in nature. However, I will outline mainstream physics and cosmology concepts as jumping off points . . . giant shoulders to stand on, so to speak.

The audience for this work is professionals in the field and everyone else with an interest in our Universe. My hope is that professional physicists and cosmologists will read this book and consider the ideas presented. Ideas are the feedstock of science, and there is plenty to chew on in these pages.

INTRODUCTION

"Science is beautiful when it makes simple explanations of phenomena or connections between different observations."

— Stephen W. Hawking (1942–2018)

Is science winning any beauty contests these days? The question is posed in light of the scientific finding that only about 4% of the Universe is made of normal matter and energy. The remaining 96% is said to be composed of mysterious dark matter and dark energy that we do not understand. As Albert Einstein said, "The most incomprehensible thing about the world is that it is comprehensible." We seem to be drifting away from that contention lately. The story to be told in this book is an attempt to get back on the trail of comprehensibility.

The tale begins in my scientific childhood, which some may say continues to this day. Here I am foreshadowing the critical reception I expect for some of the ideas to be presented in this book. Cosmology is perhaps dangerous ground for amateurs to traverse. Nevertheless, here we go.

I would like to say that the event that started me down the path to writing this book occurred 42 years ago, since of course, 42 is the answer to Life, the Universe, and Everything.[1] The event actually occurred even further back, in a grade-school class over half a century ago. It began with a science experiment and a personal experience that I apparently still remember.

---

1   Readers who have read or seen productions of Douglas Adams's excellent *The Hitchhiker's Guide to the Galaxy* will understand the reference to the number 42. For those who have not, you have not lived.

The experiment, conducted in front of the class of young science students, involved a sugar cube dissolved in one beaker of water and a spoonful of granulated sugar dissolved in another. The granulated sugar disappeared into the water faster than the cube did. The question put to the class was: Why did it take less time to dissolve the crystals than the cube of sugar?

My answer was that the energy required to break up the sugar cube into separate fragments had already been applied to the granulated sugar. Unfortunately, this was not the answer in the science curriculum that the teacher was looking for. The correct answer was that the crushed sugar had more surface area exposed to the water, resulting in faster dissolution. So, I was wrong! Or was I? I still think both answers are correct.

The lesson I learned was invaluable. I began to question whether there is only one right answer, only one correct way of looking at a problem. This question can be asked of most things in life, including politics, physics, quantum mechanics, and cosmology. In politics, one often faces the choice between conserving the past and liberating the future, for instance.

I also started thinking more about energy, the central subject of this book. Eventually, I was to have a long career in the energy industry, developing the solar energy that is stored in petroleum. Energy is life!

In essence, the purpose of the story to be told in these pages is to look at several of the major puzzles facing contemporary physics and cosmology from alternative perspectives. To see if they can be solved without recourse to mysterious explanations outside of our current understanding.

With science, proof is the crux of the matter. Is there a way to test an idea? If not, the idea is just arm-waving speculation. I have the highest regard and respect for science, and for that reason, the ideas contained in this story come with proposed experiments to test them. There are mathematical justifications presented where possible, and they can be assessed for mathematical consistency.

Whether the story turns out to be science or science fiction remains to be seen.

One test that correct ideas seem to pass is known as Occam's razor, a principle most directly stated as "the simplest answer is usually the best one." The most beautiful and profound truths are often deceptively simple. Albert Einstein's most famous equation, ($E = mc^2$), relates energy (E) to its equivalent mass (m), with the speed of light (c). How did the speed of light get in there? We shall "c" as the story unfolds.

In the absence of a multi-billion-dollar laboratory, many of the hypotheses described have been developed by the author using only thought experiments, a spreadsheet, and mathematical manipulations. Science often progresses by virtue of questions posed, leading to the generation of a hypothesis, a proposed explanation. The new idea can then be confirmed or refuted in the physical world.

Einstein was a master of thought experiments and utilized them in the development of his theory of relativity. Caution must be used, though: if a thought experiment goes wrong, it can blow your mind! That was just an experiment in humour—a concept that will be thoroughly tested throughout the book, often with dubious results.

Ultimately, the intention is to have the various ideas contained in this book reviewed by open-minded professional physicists for their critique and opinions. Many of the hypotheses described in the following pages are dramatic departures from the conventional mainstream views of how the Universe works. Some of the proposals may appear to be a recycling of old ideas that are now out of fashion. For instance, something akin to the aether,[2] a medium once thought to occupy all of space, is proposed. Wait . . . come back! I promise a completely fresh and original take on everything!

The topics to be explored are not trivial and may prove to be challenging for readers who are not familiar with the current puzzles and ongoing debates in contemporary cosmology. A number of mathematical arguments are of necessity included because math is the language of science. Those uncomfortable with the math may skip the equations and focus on the accompanying ideas and the illustrations to get a reasonable picture of the current scientific issues and the proposed new paradigm.

---

2   The aether, or ether, was a media that light waves were thought to travel through, akin to the air that allows the transmission of sound. The scientific consensus is that the aether is not required and does not exist. Any mention of the aether is usually greeted with comments such as, "The nineteenth century called and wants its physics back."

A certain amount of background research may be necessary if the reader is completely unfamiliar with the theories and tenets of relativity, quantum mechanics, and physical cosmology. A list of keywords that Google is familiar with is provided at the end of the book. The fundamentals will be explained as clearly as possible, but other sources may prove necessary to understand the topics in depth and detail. This book is a great opportunity to introduce yourself to the subjects, if they are of interest.

THE ROAD AHEAD

So where are we going with this story? First, we will look at the nature of time itself. Space-energy will be introduced as an alternative framework to space-time. Then a new and unconventional theory of gravity that differs from the conventional curved spacetime explanation will be revealed. The implications of a variable speed of light over the life of the Universe will be explored.

We will examine perplexing dark matter and dark energy with a view toward eliminating the need to use them to explain the contemporary astronomical observations that give rise to them.

Along the way there will be a new description of what forces are, an inescapable look at black holes, and a brief look at the curious world of quantum mechanics. The apparent acceleration of the expansion of the Universe will be challenged. Gravity waves and gravitational anomalies will be weighed. Oddities in the behavior of light in prisms will be viewed through a new window. A possible way to unify the four fundamental forces will come together. Possible mechanisms behind inertia and entropy will be brought to the forefront.

A visit to the Big Bang origin of the Universe and a new scenario describing its ultimate fate are also included in the tour.

We will **boldly** go places no people outside of *Star Trek* have gone before!

THE MISSION

The proposed paradigm of four-dimensional space-energy and a variable light speed encompasses the whole Universe and all its physical phenomena. Insofar as resolving current mysteries, it may provide answers that are more satisfying, than the current mainstream explanations of nature. Take gravity for example: conventional explanations range from the mysterious force that caused that hammer to drop on your foot to the curvature of four-dimensional space-time—from the unhelpful to the nearly unfathomable. A more intuitive way to look at gravity will be offered.

The path of modern science has led to mysterious dark matter and energy to account for actual observations. The assumption that these unknowns exist challenges understanding of the fundamentals of physics. Throughout this book alternatives are suggested that honour all the observations and data while offering explanations that are well within the bounds of known physics.

The new paradigm is possibly correct, possibly wrong, but definitely different and original. Most importantly, it submits to scientific testing, and therefore proof or disproof.

The mission of this book, should you decide to read it, is to connect the observations I have made about physics and cosmology over the years into a cohesive framework, as easy to take as a sugar cube. If even one of the original ideas proves useful, it will have been a worthwhile exercise.

Your view of the Universe may be changed, but if not, the book will hopefully at least be an entertaining read!

TIME

1

TIME

The following discussion is a critical examination of the concept of time. The fickle aspects of time are exposed, and its very existence, except as a useful human construct, is called into question. The argument will be made that the fundamental properties of the Universe are space and energy, with time as a secondary feature that emerges as a result of change caused by motion.

Modern physics treats time as literally the fourth dimension, in addition to the three of space. My goal will be to replace the fourth dimension of time with one of energy. Although this may seem like a waste of time, the reasons for the alternate paradigm will become clear as we progress.

Time is deeply interwoven into our lives, and there is never enough of it. Innumerable examples of people contemplating the nature of time and writing about it have occurred throughout history. Now is the place in our tale to clarify the nature of time, while continuing with the theme of simplifying things.

For our purposes, let us consider time in a specific way:

$$t = \frac{d}{v}$$

where (t) is time, (d) is distance, and (v) is velocity. Time is a ratio of distance and velocity. From this simple view of time as a ratio will come insights that threaten the existence of time as a dimension.

You may recall this relationship from school and, of course, use it in everyday life. If you decide to travel from your home to a destination 100 kilometres away, you can estimate the time it will take by considering the velocity you expect to travel at, perhaps 50 kilometres per hour, in which case it might take two hours. Or you may plan to make better time by travelling at 100 kilometers per hour and arrive in one hour.

The foregoing example illustrates the dependence of time on the energy expended to accomplish an action. Traveling at a higher velocity means more kinetic energy has been acquired. The time experienced is a simple function of the distance and energy involved. This is a bit of an inversion of the normal way of regarding time, but time is a stubbornly persistent illusion.

Three key points about the nature of time are next considered in some detail: (1) that time is entirely subjective to the observer; (2) that time is simply a ratio derived from physical space and energy, and (3) that time is a human construct for our convenience.

1. TIME – A FUNCTION OF DISTANCE AND VELOCITY

Time is just a function of distance and velocity, emerging in situations involving space and energy. When regarded in this way time becomes a dependent variable, depending on the distance and velocity involved. This is quite different than regarding time as a dimension with an independent existence.

Modern science, at least since Einstein developed the theory of relativity, recognizes the fact that time is relative. Time is known to slow down for a moving observer relative to the time measured by another observer who is stationary. The degree to which time slows down for a moving observer is directly related to the magnitude of their velocity with respect to the stationary observer. Relativity specifies an ultimate maximum velocity, a cosmic speed limit. At this limit—the speed of light in a vacuum—time is thought to stop entirely, at least relative to a frame of reference that is not moving. Time can fly by or stop completely, depending upon how much fun one is having. We will next examine how subjective it can be.

TWINS OR TRIPLETS?

You may have heard of the twin paradox, a staple of science fiction. One twin remains on Earth while the other flies off at great velocity through space and experiences time dilation as a result. At some point, the travelling twin returns to Earth to discover that her sister has aged more rapidly and is already collecting social security.

Let us expand the twin paradox to include a triplet sister. In doing this, we engage a third observer located some distance away to observe the other two. This observer is akin to us, observing the paradox from a position removed from the participants, on the outside, but in an equally valid frame of reference.

Let us give the triplets individual atomic clocks, initially synchronized, to measure time from their individual perspectives. The first sister and clock will be located at an airport. The second will be located on a jet airplane at the airport and will depart to fly around before returning to reunite with her sister on Earth. The third will be located elsewhere, perhaps on the Moon, with a powerful telescope that allows her to observe the other two. She will see the twins together, apart, and together again, and will mark the interval of time that passes for her during that sequence of events. The third sister will confirm that regardless of what the atomic clocks of the other two measure, there was a common interval that they were separated, from her vantage point.

In the theory of relativity, there is no preferred point of reference, all observers are treated equally. The laws of physics are the same for any observer in an inertial (not accelerating) frame of reference. Each of the three triplets has equal claim to their unique experience of elapsed time as being the correct one. Each sister will believe her respective elapsed time is the actual value.

There is no way to establish the correct time interval, if indeed any of them are correct. That is, unless we consider distance and velocity, within the context of energy. By looking at time as a ratio of two parameters that represent space (distance) and energy (velocity), we can perhaps find a logical path out of the paradox.

Most definitely there is an absolute physical sequence of events. Two of the triplets are together, then apart, then back together. What then is different for the three observers, each of whom regards the time interval they personally experience as correct? They each have a unique individual experience of motion through space and a unique energy consumption history.

The three sisters experience different relative motions, resulting from their different energy usage. These experiences can be distinguished, even though no frame of reference has priority. Strictly speaking, with no preferred frame of reference, the sister at the airport and the sister on the jet airplane are in motion with respect to each other, and with respect to the third sister on the Moon. From the points of view of the twins this could mean the airplane is flying around the Earth or the Earth and airport are moving around a stationary airplane.

From an energy perspective, however, the jet must be in motion around the Earth. The jet burns a quantity of jet fuel, a quantity of energy entirely insufficient to move the entire planet around the airplane. This is a critical distinction between the first two sisters: they experience a different path through space and a different amount of energy usage.

The third sister, on the Moon, can confirm that the others initially had the same gravitational potential and kinetic energy, as they were at the same location with no relative velocity with respect to each other. Then the twin on the airplane acquired kinetic and potential energy by burning jet fuel and moved relative to the one at the airport. Later, they came back together with the same energy, at the same location on the ground, with no velocity relative to each other.

This example was employed because it has really been done with individual clocks. One of a pair of synchronized atomic clocks was flown around on an airplane then reunited with its twin on the ground and their respective measured elapsed times compared. The clock in motion does run slower than the one that remains on the ground! Even more strange is the fact that a clock lifted higher in the Earth's gravitational field runs slower than another at a lower position.

How can this time discrepancy be true when a third observer can see that the time taken for the experiment, the physical together-apart-together sequence, was identical.? We find that time must be relative, not absolute, and a function of position and energy.

The differences in physical motion and energy consumed are absolute, but the time measured varies. Perhaps our methods of measuring time are being influenced by position and energy? We might also ask whether time itself is real or just an illusion?

2. TIME – A RATIO OF SPACE AND ENERGY

Time is just a ratio of two measured parameters that represent space (distance) and energy (velocity). Consider another well-known ratio we call density, defined as the ratio of mass to the volume occupied by the mass. Density does not exist independently of the mass and volume it describes. So, we may wonder whether time, as a ratio of two fundamental quantities, has an independent existence? The logical conclusion is that time does not exist independent of distance and velocity.

Think about how we measure time. You will notice that it is always by comparison to the distance and motion of other objects. A year is how long it takes the Earth to travel around the Sun in its orbit. A day is how long it takes the Earth to rotate on its axis. All clocks depend upon the motion of a pendulum, the stretch of a spring, the vibration of a crystal, or the oscillation of an atom in the case of atomic clocks, to define time. One second, our fundamental unit of time, was formally defined as 9,192,631,770 oscillations of a Cesium 133 atom by the General Conference on Weights and Measures back in 1967, many Earth orbits ago.

Imagine yourself as a consciousness in an empty Universe with no reference to calibrate distance or motion. How would you perceive time? We can get some idea from reported experiences in sensory deprivation tanks. These tanks attempt to remove all environmental stimulation by being soundproof, dark inside, and floating the occupant in body temperature salt water, their phone having been collected at the door. People who spend time in sensory deprivation tanks typically lose all track of how much time has elapsed.

3. TIME — A CONVENIENCE

And the third point: Time can be argued to be a convenient human construct. Time is only an agreed upon reference to coordinate our activities—a device for scheduling events. Attention time travellers, meet here last Thursday at 7:00 p.m.!

SPACE AND TIME

The assertion that time does not exist has a drastic implication for space-time, the four-dimensional Universe that we are thought to be living in. Space-time is constructed from three dimensions of space and one dimension of time. If we can ask whether time is real, should we also be able to ask whether the time component of space-time is real?

We can probably agree that space is real. One can easily visualize space, described as three orthogonal (at 90 degrees to each other) dimensions, by considering how many coordinates are needed to specify the location of any point in space. The Cartesian coordinates (x, y, z), typically used for three-dimensional representations of length, width, and height often encountered in daily life, are familiar, as represented in Figure 1.1. The time coordinate is slippery, though—for one thing, it keeps changing.

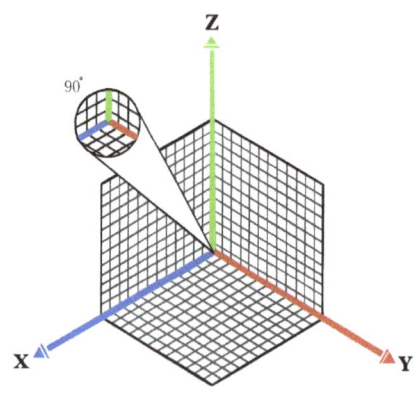

Figure 1.1. 3D Space

The goal of this book is to frame everything about the Universe in four dimensions: the three of space, and one of energy, instead of time. Doing so has important advantages, including a new theory of gravity. Time keeps flowing like a river into the future . . . but rivers really flow toward lower potential energy.

Energy is less elusive than time in one extremely important way: energy is conserved, but time is apparently not. As we saw with the triplet paradox, an energy audit can distinguish between frames of reference.

ENERGY

Energy comes in a variety of interchangeable forms. Every object has a rest mass when it is not in motion, theoretically convertible to energy by the previously mentioned ($E = mc^2$). The object may also have a temperature, correlated to the rapidity of motion of its constituent atoms. A kinetic energy dependent on its relative motion or velocity may also be present. In addition, the object can have gravitational potential energy dependent on its distance from another mass.

Energy can be neither created or nor destroyed, but we can make time or waste it. All joking aside, there is a good reason for this alternative philosophical approach and the space-energy model.

SPACE-ENERGY

We have seen that time depends on energy, and that should give the latter a firmer foundation in reality and a more fundamental importance. The triplet paradox showed that two frames of reference can be distinguished from one another if energy consumption is accounted for, and that time flexes with changes in kinetic and potential energy.

Throughout the book, we will see how the concept of space-energy comes into play. The title of the book is a play on words (The Forth Dimension = The Fourth Dimension = Energy). Much more than just playful fun though, this perspective provides an alternative explanation of gravity that is more intuitive than curved space-time.

How then does the gravity hypothesis in this book relate to gravity as explained by Einstein's theory of relativity? Einstein's relativity has been tested in many circumstances and proven correct in every case. Therefore, it would be ridiculous to come up with an alternative theory that was inconsistent with relativity. That is not the objective of the current effort.

Instead, the question raised is whether relativity can be reformulated in a space-energy context rather than in a space-time context. We will see that such reformulation is possible with one of Newton's laws of motion.

The distribution of mass and energy in the Universe, rather than influencing the curvature of space-time, would instead determine the shape of an energy dimension, the intensity contours of an energy background. There are solid mathematical foundations for this view that will soon be shown in detail.

Unfortunately, attempting to restate Einstein's field equations in terms of energy is beyond the scope of this author, but might be attempted by someone with a black belt in tensor calculus. However, it can be said that the alternative space-energy view is supported by certain other mathematically consistent logical propositions. These observations lead to very interesting conclusions that will next be revealed.

A brief word about relativity: I had a favourite aunt who lived in a nice part of the world and was fun to visit; that is special relativity. The child of your father's cousin is your second cousin; that is general relativity. Neither of these theories are Einstein's, but the theory of this book is to have a bit of fun!

SPACE-ENERGY

2

"Nothing exists except atoms and empty space; everything else is opinion."

— Democritus (c 460–c 370 BC)

Figure 2.1. Stargazing

Gaze up at the stars on a clear night. Pick out a particularly bright star and keep your eye on it as you walk around a little. You will continuously see it up there in the sky. The light from that star must be falling all around you, but it is only the light falling on your eye that you are aware of.

Now find another star so that you have two within your view. Light from both stars must be hitting your eyes. Suppose you were near one of those stars, looking at the other. You would see the light from the other star. Light must be travelling from one star to the other as well as to Earth. The star-to-star, interstellar, light is not observable to the ground-based observer, as it does not fall on a direct line to their eye.

The idea becoming apparent from this starry night experiment is that space is not empty at all—it is full of light! Sorry Democritus, that is my opinion, and despite your name, you do not get a vote. Energy is continuously radiating out in every direction from celestial objects as rays of light.

Light is composed of electromagnetic waves or the equivalent massless particles we call photons. Wave-particle duality is a concept from quantum mechanics that we will put up on the hoist in a later chapter, to inspect its curious workings.

For now, the key idea is that space is full of energy. That energy is comprised of not only the rainbow of visible light, but a full range of all the frequencies on the electromagnetic spectrum from long, low-frequency radio waves to short, high-frequency gamma radiation. Figure 2.2 shows the full spectrum. The electromagnetic energy is there even if it does not meet the eye. Even if the eye does not respond to it, as is the case for frequencies outside the visible range, radiation is all around us.

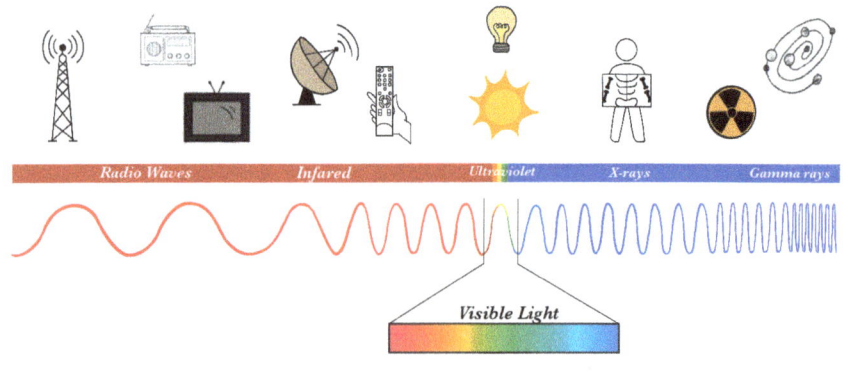

Figure 2.2. The electromagnetic spectrum

The energy background filling the Universe is the fourth dimension in this thesis. There are the regular three dimensions of space, plus an energy dimension. The magnitude of the energy dimension is the energy intensity or density, the amount of energy within a given volume of space.

A demonstration of a kind of four-dimensional space-energy is available wherever you may be reading this. Let us assume you are indoors, in a room. You can specify any point in the room with three coordinates relative to an arbitrary origin. Say you choose a spot on the floor, then determine that the point you are interested in is three feet to the left, four feet forward, and six feet upwards (better than six feet under). You can measure the temperature at that point in the room, locating its coordinate in a fourth dimension of thermal energy. The temperature is a measure of the average velocity of the air molecules, the kinetic energy, of the air at that point.

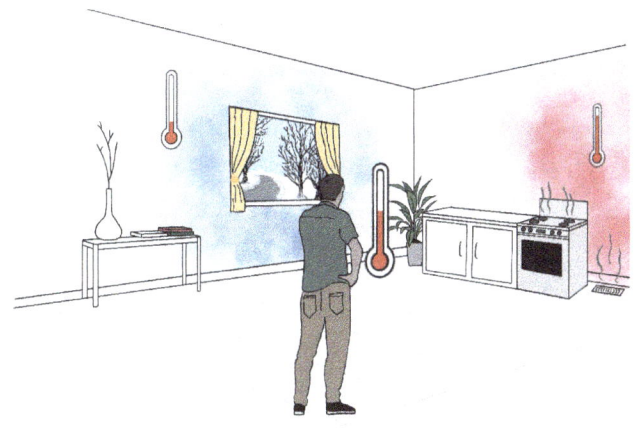

Figure 2.3. The temperature dimension.

Measurements of temperature at various points in the room would reveal a varying energy distribution throughout the room, perhaps warmer around the lights, cooler around the window, hot above the stove. Those measurements of energy intensity could be plotted and contoured in space. One could also measure the time at each point in the room if one were interested in mapping out space-time, but I digress.

With changing energy values from point to point come energy gradients. We will use a shorthand mathematical label for an energy gradient, dE/dx, a small change in energy corresponding to a small change in distance. For instance, an average down-ward slope in temperature from a point above the stove to a point beside the window is the change in temperature divided by the change in distance between the locations.

Like fish in the depths of the ocean feel no water pressure because there is an equal pressure within their bodies, we are not aware of the background energy, except in relative terms. With respect to the temperature in the room, our bodies try to maintain a constant temperature, so we may notice if the room is comparatively warm or cold, merely affecting our comfort. If a deep-sea fish is brought to the surface, the reduction in pressure is a big problem for it!

Energy is in fact within us, in surprising quantities. According to Einstein's relationship ($E = mc^2$), mass and energy are two different forms or phases of the same thing. A comparison could be made with water vapour and ice; they are two phases of the water molecule. One kilogram of mass, multiplied by the speed of light squared, contains roughly 9×10^{16} joules, roughly the energy in 14.5 million barrels of oil. People on average have a mass of 62 kg, and an equivalent energy of roughly 900 million barrels of oil. The world uses about 100 million barrels of oil per day.

This energy equivalent of our body mass cannot be released unless we encounter an antimatter version of ourselves and annihilate it, but theoretically, it is present. We contain a significant amount of energy, in the form of mass, even if we do not feel it. Even if we are tired at the end of the day!

THE UNIVERSE IN A GLASS OF WATER

Our next small step is one giant leap, to consider the entire Universe. The Universe has a three-dimensional volume we call space. Space contains mass and energy, interchangeable in tiny fractions within thermonuclear devices such as stars, or in the spontaneous radioactive decay of some elements. Stars are effectively nature's mass to energy converters, and the Universe is filled with fusion reactors.

Later, we will explore a possible explanation of how mass and energy are two phases of the same thing. For now, let us consider a glass of ice water, represent-ing the Universe, in a thought experiment.

The glass of ice water has a volume and contains ice and water, two forms or phases of the same H_2O molecule. The ice is slightly less dense than water, and it floats mostly immersed in the water but with the proverbial tip of the iceberg above the surface. Ignore that minor difference in density for a moment. Measuring from the surface of the water to the bottom of the glass, we find the water has a certain depth. Where there is an ice cube, we measure down from the surface a certain thickness of ice, then continue from the bottom of the ice to the bottom of the glass through a reduced column of water, as illustrated in Figure 2.4.

Figure 2.4. The Universe in a glass of water

At places in the glass where there is ice floating, the thickness of water beneath the ice is less than the total water depth where there is no ice. Overall, the glass of water maintains a level surface (we are ignoring the fact that the ice sticks up a little bit above the surface for this illustration). If we move an ice cube sideways with our finger, the water rapidly fills in around it, maintaining a level surface.

By analogy, the glass is the Universe, the ice is mass, and the water is energy. Stars might be figuratively thought of as bits of ice, melting energy into the Universe, although that is only a very simplistic representation. Stars operate through nuclear fusion, of course, not by melting. The water level in the glass is akin to the average mass-energy density of the Universe. One might also ask whether the Universe is half full or half empty, but we won't go there!

The most important takeaways from the glass of ice-water model: (1) where mass (ice) is present, there is less energy (water) present; and (2) the depth of the water representing the energy phase varies where mass is present, causing energy gradients. These facts are crucial to understanding the alternative mechanism of gravitation being proposed.

Another simple model that is frequently used in discussions about mass and its gravitational effect on space-time is a bowling ball on a rubber sheet. We are often constrained by the need to describe a four-dimensional phenomenon in terms of three-dimensional objects familiar in daily life. The bowling ball deforms the rubber sheet, creating a curved depression, and other objects, say small balls, will roll into the local low area. In 3D space the gravity well of a spherical object such as a star is a spherical shape.

In relativistic terms, the mass causes space-time to curve, and the balls follow lines called geodesics representing the shortest distance through curved space-time. Notably though, the energy content of the small ball as it rolls determines the line it will follow. A high-energy ball might pass right by with only a slight deflection, and a low-energy ball might go into orbit or fall into the large ball.

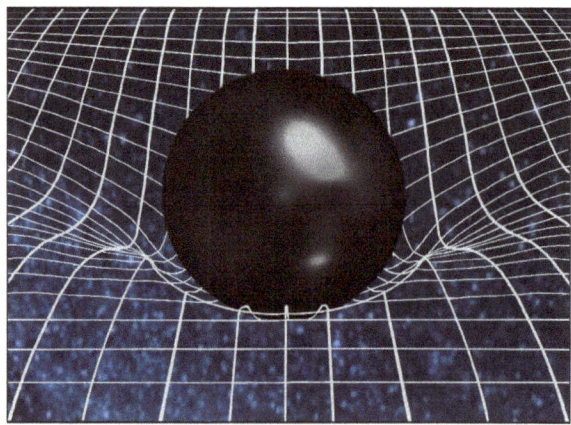

Figure 2.5. Rubber Sheet Analogy

In the space-energy scenario the background energy is displaced by the mass, just as the ice cube displaces water in the glass. That is the concept of a contoured energy dimension, and it turns out there is mathematical support for this view.

More will be said later about Einstein's field equations that relate the distribution of mass and energy in space to the curvature of space-time. For now, the question is posed: Does it make sense that mass causes a curvature or contouring of the background energy? And do slopes or gradients in the energy background result in forces?

You may be beginning to realize the "gravity" of this situation . . .

GRAVITY

3

"What is there in places almost empty of matter and whence is it that the sun and planets gravitate towards one another, without dense matter between them?"

— Isaac Newton (1643–1727)

Gravity is a puzzle. Sir Isaac Newton, who developed the law that quantifies the effect of gravity, could not explain precisely how it works. With the apparent demise of the aether, a media thought to permeate the entire Universe and serve as substance for the propagation of light waves, there is indeed nothing left between gravitationally attracting objects.

Enter Einstein and his theory of general relativity, explaining gravity in terms of curvature of the four-dimensional fabric of space-time itself. The curvature is a result of the distribution of mass and energy throughout space. We are familiar with three-dimensional space (length, width, height) and the passage of time (yesterday-tomorrow). Objects are now said to follow geodesic paths through curved space-time, giving the appearance of a gravitational force in action when in fact there is no force. Geodesics are the shortest possible course between two points on a curved surface. Airliners tend to follow geodesics or great circle routes to minimize travel distance around the Earth.

The unification of the four fundamental forces of nature (gravity, electromagnetic forces, and the weak and strong nuclear forces) is one of the major goals of physics, but it remains elusive, at least until later in this book. Gravity acts on the cosmic scale while quantum mechanics describes the behavior of particles on the subatomic scale. The quest of physicists since these two great theories of the twentieth century were conceived has been to unite them. One possibility for accomplishing the unification will be discussed in a later chapter. A different concept of what gravity is, and what forces in general are, is required first.

GRAVITY

Gravity has the associated concepts of orbital speed (v_o), and escape velocity (v_e). Figure 3.1 is the often-used depiction called Newton's cannonball. A cannon shoots a ball at ever-increasing velocity, so that it lands further and further away. Ultimately, the ball falls all the way around the Earth, achieving orbit, or with even greater velocity, it leaves the Earth entirely.

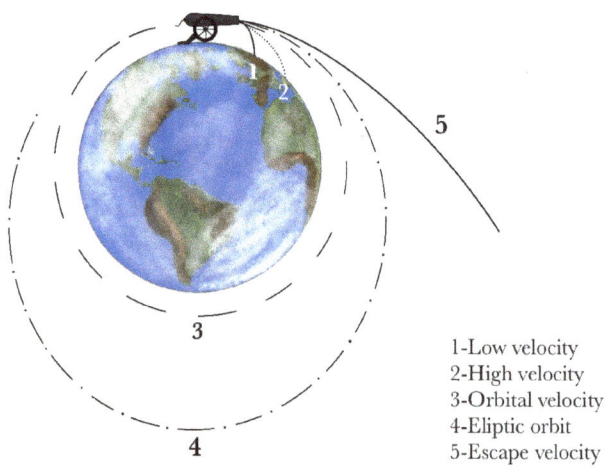

1-Low velocity
2-High velocity
3-Orbital velocity
4-Eliptic orbit
5-Escape velocity

Figure 3.1. Newton's cannonball

The orbital speed (v_o) is the speed an object must have to orbit another object of much greater mass when the two are gravitationally bound. This relationship applies to satellites in Earth's orbit and planets orbiting stars.

The orbital velocity is given by,

$$v_o = \sqrt{\frac{GM}{r}}$$

where (G) is Newton's gravitational constant, (M) is the mass the satellite is orbiting, and (r) is the radius of the orbit.

An object needs to possess an even greater speed to escape from the gravity of another object. That is the escape velocity, the least speed required not to be attracted back by gravity. The escape velocity (v_e) is the speed at which the sum of an object's kinetic energy (E_k) and gravitational potential energy (E_p) is zero.

$$E_k + E_p = 0$$

$$\frac{1}{2}mv^2 - \frac{GMm}{r} = 0$$

$$v_e^2 = \frac{2GM}{r}$$

$$v_e = \sqrt{\frac{2GM}{r}}$$

Escape velocity: you cannot leave your home world without it!

For the Earth, escape velocity works out to 11,186 metres per second. For much more massive Jupiter, it is 59,500 m/s. For a black hole, the escape velocity is the speed of light, 299,792,458 m/s. Black holes will be discussed in an inescapable way in a moment.

Curiously, a factor of the square root of two is the difference between orbiting and leaving altogether.

$$v_o = \frac{v_e}{\sqrt{2}}$$

There are many curious things that result from the mathematical descriptions of nature, and this is yet another one. If we consider a unit vector, an arrow of length one, and another unit vector orthogonal to it, their sum has the length

$\sqrt{2}$. That is just the way it is, but is it a clue to something fundamental about the Universe? Not sure, just an observation.

Note that the behaviour of objects under the influence of gravity depends on their velocity, and by extension, upon their kinetic energy content per unit mass, or their "specific kinetic energy" content.

$$\text{Specific Kinetic Energy} = \frac{E_k}{m} = \frac{1}{2}v^2$$

The postulated energy background, as the fourth dimension, is the basis for the new theory of gravitation. The background energy is imagined having an energy density variation with contours like the hills and valleys of a rolling landscape. Areas of low energy are caused by the presence of mass, in much the same way that ice cubes in our "Universe in a glass of water" experiment result in areas of less water depth. To state it most simply and directly, gravity is caused by gradients in the energy background. In a later chapter, we will see that all forces are energy gradients.

In the analogy of a landscape with hills and valleys, a two-dimensional surface like the surface of the Earth is invoked. In the three dimensions of space, the energy valleys are spherical in shape, making visualization much more difficult.

Discrete bodies in space shaped by gravity are generally approximately spherical, like stars and planets. Spherical is the lowest energy configuration, but real bodies can be deformed if they are spinning. Spinning objects have another form of energy called rotational energy or angular kinetic energy. Because the Earth spins, it deforms into an oblate spheroid, with a larger diameter at the equator than from pole to pole.

MEASURING THE ENERGY BACKGROUND

The profile of the energy background can be calculated mathematically. The intensity can also be mapped out experimentally, and we humans are doing it more often now than ever before. The very first time we probed the contours of the energy dimension outside the Earth's atmosphere was on October 4, 1957. The then Soviet Union launched a satellite called Sputnik 1 into orbit.

In orbit, the force of gravity disappears. Why?

A satellite, if it is to stay in orbit, requires a certain energy. First, it must be lifted to the height of the orbit above the Earth to give it the required potential energy to be at that altitude. Lift your coffee cup off the desk. This requires an energy input to overcome the force of gravity over the distance you lift the cup. Force multiplied by distance equals work, another name for energy. Even if you are having a coffee, you are still working!

However, the satellite needs something else to stay in orbit or it would simply plummet back to Earth. A satellite needs orbital velocity (v_0), a precise amount of kinetic energy. How is the energy necessary to orbit the Earth acquired? Normally, by converting the energy contained in a chemical fuel to potential and kinetic energy, using a rocket.

The combined potential and kinetic energy needed for a satellite to go into orbit at a certain altitude indicates the magnitude of the energy background at that altitude above the mass of the Earth. Where there is no energy gradient or difference in the specific energy the object has and the background, there is no force. The satellite feels no gravitation, just as people aboard the International Space Station still have mass but experience weightlessness.

A NEW PARADIGM FOR GRAVITY

The foregoing is not the conventional explanation for gravity. The accepted explanation is that the mass of the Earth causes local space-time to curve, and the satellite merely follows geodesic lines through curved space-time. That is one way of looking at it, but is it the only way?

The space-energy theory proposed as an alternative to curved space-time makes a prediction that can be confirmed mathematically, much to the author's amazement! The prediction is that the energy equivalent of any mass, as derived with ($E=mc^2$), should be the amount of energy required to figuratively fill in the gravity well caused by that mass. This turns out to be true. The strongest underpinning to the space-energy hypothesis is actual mathematical proof!

The best way to understand how energy and gravity are related is to imagine taking a body such as the Earth apart bit by bit. To remove a grain of sand from the Earth and launch it right out of the gravity well of the planet to an effectively infinite distance away requires a certain amount of potential energy be added to the sand grain. Essentially, it must be lifted from the surface of the Earth to a height approaching infinite. Mathematically, the distance is infinity, but for practical purposes, a large distance, such as to the edge of the Universe, will suffice.

Now imagine repeating this process, grain by grain, until the entire Earth is dispersed to infinity. That total amount of energy input needed is equal to the energy content of the mass of the Earth as calculated by applying ($E = mc^2$), where (m) is the total mass of the Earth. The mathematical proof of this astonishing conclusion will next be outlined in detail.

Newton not only developed the law of gravity, but he invented the calculus as well. Integral calculus is a mathematical technique that allows the adding up of a huge number of infinitesimal parts to arrive at a total. In our case, integrating all the grains of sand that make up the planet Earth would give us the total mass of Earth.

Newton's law of gravity is given by the equation,

$$F = \frac{GMm}{r^2}$$

It turns out that integrating the force of gravity over distance tells us the amount of gravitational potential energy required to lift a grain of sand from the Earth to an infinite distance away. Logically, removal of the object also removes its gravity well, its effect on the local curvature of space-time, or its associated energy background deficit.

The equation for gravitational potential energy (U) and its derivation by integrating the force of gravity appears below. (G) is the gravitational constant, (M) is the mass of the Earth in this case, (m) is the mass of the grain of sand, and (r) is the distance from one to the other. The symbol $\int$, which looks like a stretched-out letter "s," is the symbol for integration, meaning to add up the force of gravity times the distance, from a distance r to infinity in little increments of distance (dr).

$$U = \int_R^\infty \frac{GMm}{r^2} = -\frac{GMm}{r} \Big|_R^\infty$$

This is the height of the complicated math that will be encountered in this book, so do not despair—it is all downhill from here!

How do we choose a distance (r) as the lower limit of integration? We cannot integrate from zero radius, as it causes infinity to arise if zero appears in the denominator of the integrand. As we remove grains of sand from the Earth, it gets smaller, so the radius changes as we go.

The discovery the author has made (at least for himself—I am not aware of this being done elsewhere) is that it is necessary to integrate from a particular radius to infinity. That special radius is called the Schwarzschild radius, or the gravitational radius. Every mass (M) has a characteristic Schwarzschild radius as given by the equation below. The terms that show up are the same gravitational constant (G) and mass (M) and speed of light (c).

$$r_s = \frac{2GM}{c^2}$$

Using the integral above and applying the Schwarzschild radius (r_s) for R, and recognizing that when (r) is infinity the potential energy expression is zero, it turns out by the rules of calculus to be,

$$U = \left[\frac{-GMm}{2GM/c^2} - 0\right] = \frac{1}{2}mc^2$$

On average during the process, because the Schwarzschild radius is proportional to the mass,

$$r_s = \frac{2GM}{c^2} \times \frac{1}{2}$$

Therefore, our integral becomes,

$$U = \left[\frac{-GMm}{GM/c^2} - 0\right] = mc^2$$

Alternatively, we could assume the Earth is first divided in two equal pieces and separated so that the operation needs to be repeated twice, once for each piece.

We arrive at the rather astonishing conclusion that if a mass such as the Earth was converted to its equivalent energy according to ($E = mc^2$), it would exactly fill in the energy missing from its gravity well, the depression in the background energy caused by the mass. The gravity well would vanish as the mass was removed.

You have seen the typical illustration of a ball on a rubber sheet that shows by analogy how a massive object warps or curves space-time. Figure 3.2 is a déjà vu appearance of it.

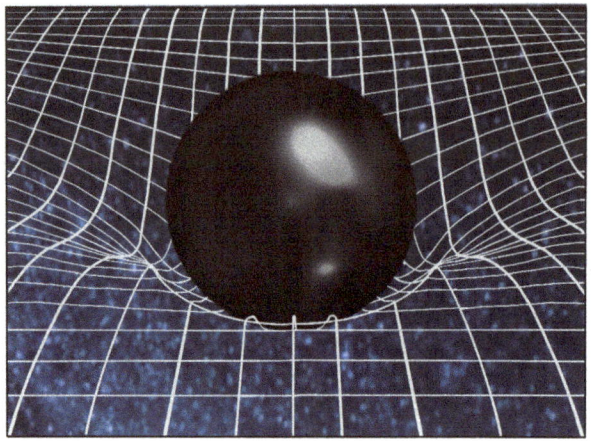

Figure 3.2. Rubber sheet analogy again

The new way of looking at this being advocated here, is that the mass is really causing a depression or deficit in the energy background. By taking the sphere pictured apart piece by piece and removing it, the rubber sheet would return to a flat state and no gravitational force would be felt because the energy background would be uniform (zero slope or zero gradient) throughout space.

THE BALLOONIVERSE

Consider another analogy to expand upon the idea, if you will, a balloon. Just like the air pressure inside a balloon is the same everywhere inside the balloon or becomes the same (equilibrates) if disturbed, the mass-energy density inside the Universe tries to maintain an equilibrium. The fabric of the Universe has imprinted upon it a pattern governed by the configuration of the interwoven mass-energy. This is akin to striving for and maintaining thermodynamic equilibrium but is in fact a gravitational mass-energy equilibrium.

The Universe, like a balloon, is expanding. We have determined that galaxies are moving away from us in all directions and moving faster at greater distances, an observation consistent with expansion of the Universe. Over the life of the Universe since the Big Bang, the energy and mass density (ratio of mass or energy to volume) have been changing. The change in mass-energy density has significant implications for the speed of light.

THE STORY SO FAR

A quick review of the new model up to this point is appropriate.

First, time is recast in the role of simply and only a ratio of distance to velocity. In the same way density does not exist independently of a mass and a volume. Time has lost its independent existence. Rather than looking at time as a dimension unto itself, time becomes a ratio of a space parameter (distance) and an energy parameter (velocity). We reduced time to a convenient comparison of motions for humans to coordinate their activities.

In this chapter, we reinvented gravity as a force resulting from an energy gradient, caused by the presence of mass, in the energy background. We found that an object must have a specific amount of potential and kinetic energy to counter gravity in order to orbit the Earth and experience weightlessness. When there is no energy difference between an object and the background the energy gradient is zero.

Mathematically, we proved that the energy equivalent of the mass ($E = mc^2$) is precisely the amount required to disperse the mass and eliminate its gravitational well. This is a significant fact, and possibly a new discovery.

Perhaps it is easier to imagine space as a volume full of energy with superimposed bubbles of reduced energy density around concentrations of energy locked in masses than to imagine the curved space-time of relativity theory? For me, the space-energy scenario is physically equivalent, but makes more intuitive sense, and we will continue to build the case for it.

FOURTH DIMENSION 4

THE FOURTH DIMENSION

Normally when physicists or cosmologists speak of four dimensions, they mean length, width, height, and time. The three dimensions of space are orthogonal, meaning at right angles to, or perpendicular to, each other.

Let us now investigate a remarkably interesting relationship between an ancient equation and a relatively modern equation. The two equations in question are the Pythagorean theorem and the Lorentz transformation. The purpose of the investigation is to link the three space dimensions to a fourth of energy.

Named after the Greek philosopher Pythagoras, who lived from about 570 to 495 BC, the Pythagorean theorem is a relationship between the length of the sides of a certain type of triangle. The triangle in question is the right triangle, having two sides that are orthogonal, meeting at an angle of 90 degrees. The theorem deals with lengths in two orthogonal dimensions and the length of a third segment that has endpoints along the two dimensions. Pythagoras states that the square of the length of the side opposite the right angle, the hypotenuse, is equal to the sum of the squares of the lengths of the two sides adjacent to the right angle. This formula should be familiar to most readers.

$$C^2=A^2+B^2$$

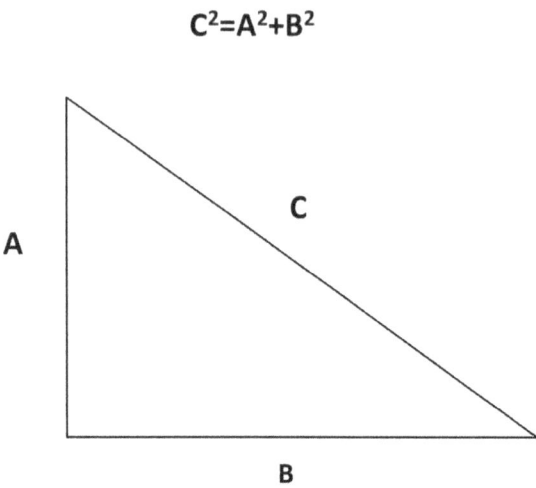

Figure 4.1. Pythagorean theorem

Less familiar will be the Lorentz transformation, developed by the Dutch physicist Hendrik Lorentz (1853–1928). Lorentz was a collaborator with Einstein on the special theory of relativity and had derived the Lorentz transformation prior to the publication of Einstein's theory.

The Lorentz transformation relates the reference frames of two observers in motion relative to one another. This allows comparison of the lengths, times, and masses in a frame of reference that is in motion relative to the observer's frame of reference.

In relativity theory, the Lorentz transformation describes length contraction, time dilation, and relativistic mass. For those not familiar with the relativistic effects of moving at high velocity, they can be summarized as follows: an object's length becomes shorter in the direction of motion; time runs slower for a moving object than for one at rest relative to it; and the relativistic mass of an object increases with increasing velocity due to accumulating more energy. The formulas below show how these things vary with velocity in relation to the speed of light.

$$L = L_0\sqrt{1 - \frac{v^2}{c^2}}$$

$$\Delta t' = \frac{\Delta t}{\sqrt{1 - \frac{v^2}{c^2}}}$$

$$m_{rel} = \frac{m}{\sqrt{1 - \frac{v^2}{c^2}}}$$

In these equations, (v) is the velocity of an object relative to a resting frame of reference, and (c) is the speed of light. The proper length (L_o) of an object measured in its own rest frame is related to the contracted length (L) measured when the object is in motion. Similarly, time is found to slow down (t') relative to its interval (t) in an object's rest frame. And the relativistic mass (m_{rel}) is found to be greater than its rest mass (m) due to the additional relative velocity, and therefore kinetic energy, of a moving mass.

The scope of this book does not permit a comprehensive review of special and general relativity, but there are many books and online resources available. A basic understanding of the subject may be necessary to fully appreciate the material to be discussed next, but please continue reading as what follows is a straightforward observation.

LORENTZ AND PYTHAGORAS

The common element in the three formulas above is called the Lorentz factor, γ.

$$\gamma = \frac{1}{\sqrt{1 - \frac{v^2}{c^2}}}$$

The Lorentz factor includes the speed of light (c), and the velocity (v) of the object that is moving. Note that if the velocity is zero, the Lorentz factor reduces to 1, and if the velocity is equal to the speed of light, the denominator becomes zero and the Lorentz factor asymptotically approaches infinity.

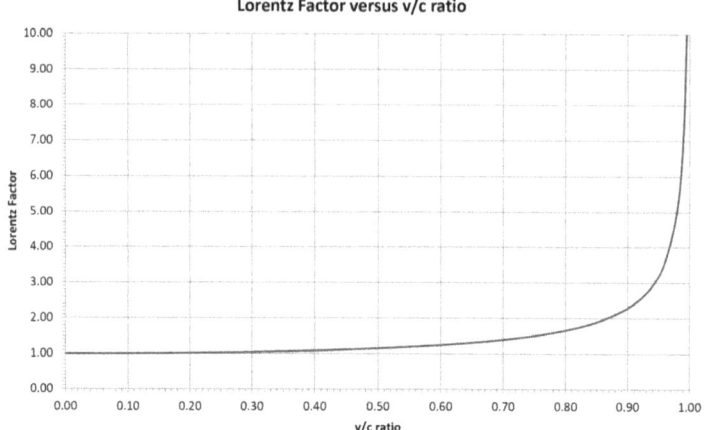

Figure 4.2. Lorentz factor versus v/c ratio

The observation I wish to make, though, is that the formulas shown above bear a striking relationship to a rearrangement of the Pythagorean theorem. This fact has always made me wonder about whether the Lorentz transformation is describing another dimension. First, let us see how this is true, and then discuss its implications.

Rearranging the length contraction equation slightly to this form:

$$\frac{L}{L_0} = \sqrt{1 - \frac{v^2}{c^2}}$$

And then to this:

$$\left(\frac{L}{L_0}\right)^2 = \left(1 - \frac{v^2}{c^2}\right)$$

And, finally, to

$$\left(\frac{v}{c}\right)^2 + \left(\frac{L}{L_0}\right)^2 = 1$$

If the sides of the right triangle are $a = \dfrac{v}{c}$, $b = \dfrac{L}{L_0}$, and $c = 1$, we obtain the familiar $a^2 + b^2 = c^2$ of Pythagoras.

What does the corresponding triangle look like?

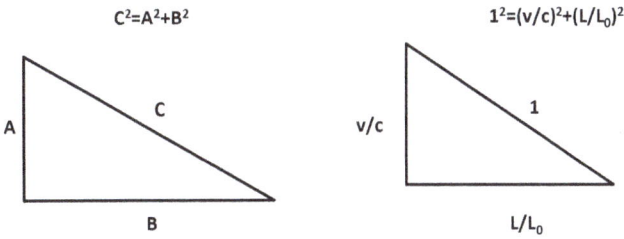

Figure 4.3. Pythagoras-Lorentz transformation

ROTATION INTO THE ENERGY DIMENSION

Suppose the hypotenuse—side C, equal to 1 on Figure 4.3 —is taken as the actual length, time, and rest mass of an object when stationary relative to the observer. Lay this quantity on the x-axis. Let us define the x-axis as one of the three space dimensions and define the y-axis as the energy dimension. Figure 4.4 depicts this initial situation.

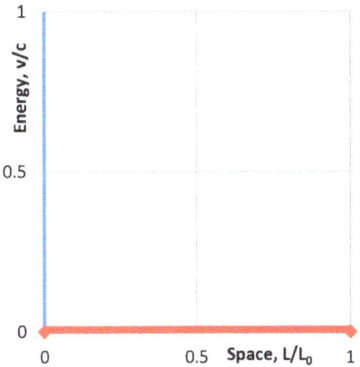

Figure 4.4. L/L_0 versus v/c with $L/L_0 = 1$

Next, move the end of the line at the origin up the y-axis, denoted the (*v/c*) axis, rotating it as we go. This is seen in Figure 4.5. The length of the line always remains as 1, as it would seem to an observer travelling with the object. However, note that the length of the object projected onto the space axis becomes shortened.

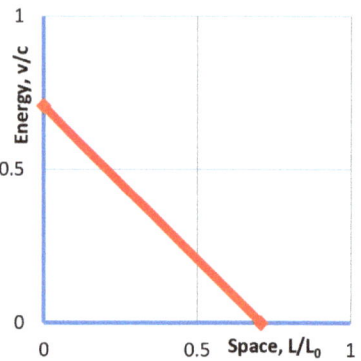

Figure 4.5. L/L_o versus v/c with $L/L_o < 1$

The projection of the red line onto the x-axis dimensions contracts until it disappears entirely when the line is vertical and aligned with the y-axis. Figure 4.6 illustrates this situation. The object is now travelling at the speed of light (*c*), and (*v/c* = *c/c* = *1*). A physical mass cannot reach the speed of light because the energy input required to do so is infinite.

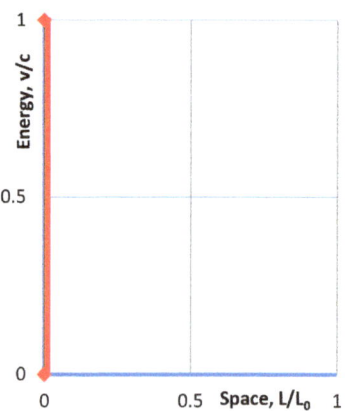

Figure 4.6. L/L_o versus v/c with v/c = 1

Photons, the particle form of light, have no mass and seem to travel in an orthogonal dimension to the three of space—or at least, that can be a useful way of thinking about it.

Note that we have been increasing the velocity of an object with mass, and hence its kinetic energy, in the process of rotating it by 90 degrees. This is why the fourth dimension is identified as being energy related.

We have found the fourth dimension!

RELATIONSHIPS OF VELOCITY, MASS, AND ENERGY

The quantity on the y-axis of the preceding diagrams is velocity/speed of light. Velocity is a proxy for energy since energy must be applied to cause velocity to increase. Velocity relates to kinetic energy (E_k) in classical physics according to the formula,

$$v = \sqrt{\frac{2E_k}{m}}$$

This only applies at low velocities relative to light, but the equation has a similarity to the relationship of energy, mass, and the speed of light:

$$c = \sqrt{\frac{E}{m}}$$

Einstein had the insight that the speed of light is the same for all observers. That means that someone travelling on a beam of light and measuring the speed of an oncoming beam of light would find it to be the speed of light, not twice the speed of light as seems intuitively correct. More will be said about how that strange effect can happen in a four-dimensional space-energy framework in a moment.

The intuitive leap proposed now, though, is to take this demonstration literally—to accept that the object as it gains energy is rotating into an energy dimension, orthogonal to the three ordinary dimensions of space. From this perspective, the fourth dimension becomes energy rather than time. Interesting consequences follow when light and gravity are contemplated, utilizing this alternative paradigm. Electromagnetic waves travelling at the speed of light ($v/c = c/c = 1$) can, in a sense, be travelling within this fourth dimension.

VISUALIZING A FOURTH DIMENSION

Perhaps you are doubtful, but in fact, it is easy to visualize a fourth dimension that is energy. We used the temperature variation in a room to illustrate the presence of an energy dimension earlier.

Another way to easily visualize four dimensions involves a sphere and a length of string. A bit of tape and a protractor is helpful too. This is not exactly the string theory you've probably heard about, but you can try this one at home. For your sphere, a globe of the Earth is ideal, with lines of longitude, an equator and poles marked on it, but any roughly spherical ball will work. I will refer to the picture of a Styrofoam hemisphere shown in Figure 4.7 to describe how this experiment goes.

Take your long piece of string and hold one end to the top of the sphere, the North Pole if you are using a globe of the Earth to do this. Run the string down to the equator and mark that length. Fold the string along that length two more times, so you have three lengths all the same. You can form these strings into an equilateral triangle with three 60-degree angles and sides of equal length, if you lay them on a flat table. Consider each string to be a dimension.

The three sides of the equilateral triangle of string, representing the three space dimensions, are not orthogonal, not at 90 degrees to each other, since they meet at 60-degree angles. They can be made orthogonal by taking the string to the sphere or globe and fixing the end to the North Pole. Then run the string down to the equator and fix it at the point where they cross. Then run the string along the equator for the same length and attach it, and then take it back up to the North Pole and tape it there.

Figure 4.7. Four orthogonal dimensions

Now the three dimensions are at 90 degrees to each other, as can be verified with a protractor. The strings lie on the surface of the globe representing the surface of the Earth. Incidentally, "Flat Earthers" will still have 60-degree angles in their string. Oh well, what can you do?

Now, what about the fourth dimension? Directionally, it is simply any line running through the centre of the globe and exiting perpendicular to its surface. This fourth direction is orthogonal to the other three. In the photo in Figure 4.7, a knitting needle has been inserted at the top to delineate the fourth dimension.

The globe and string example is just to demonstrate that four orthogonal dimensions are conceivable without constructing a tesseract as does the protagonist in Robert A. Heinlein's 1941 short story "And He Built a Crooked House"

The four-dimensional sphere model can be used to explain how light always travels at the speed of light, regardless of the velocity of the observer in relation to it. This is not intuitive, since we regularly experience objects moving toward each other at a closing velocity that is the sum of their individual velocities. Imagine two Eagles in cars, approaching each other on a dark desert highway, cool wind in their hair . . . Well anyway, if approaching each other at 100 kilometres per hour, they will have a relative closing velocity of 200 km/hr.

An observer on a light beam, though, would find an oncoming light beam to be approaching only at the speed of light. How can that be? The four-dimensional space-energy model provides a workable explanation.

We will designate the string that runs on the equator of the globe, horizontally along the bottom of the triangle of strings in Figure 4.7, as a space dimension. We will designate the two vertical strings going up to the top as the orthogonal energy dimension. Then we will imagine two photons are launched, one up each vertical string.

The difficult part of this concept is now introduced. The two photons in the energy dimension are projected on the space dimension with lines orthogonal to their trajectories. We will say it takes the photons one second to traverse the length of the strings and meet at the top, the North Pole. After half a second the photons are halfway up the strings, and their orthogonal projections onto the space string at the bottom have met in the middle. After one second, the projections of the photons have switched sides along the bottom string.

An observer in the space dimension would see the photons cross and switch positions in one second, and their closing velocity would obviously be twice the speed of light. An observer on either of the photons, though, would see the other closing at only the speed of light, and they would meet up in one second. Figure 4.8 is a diagram of this thought experiment.

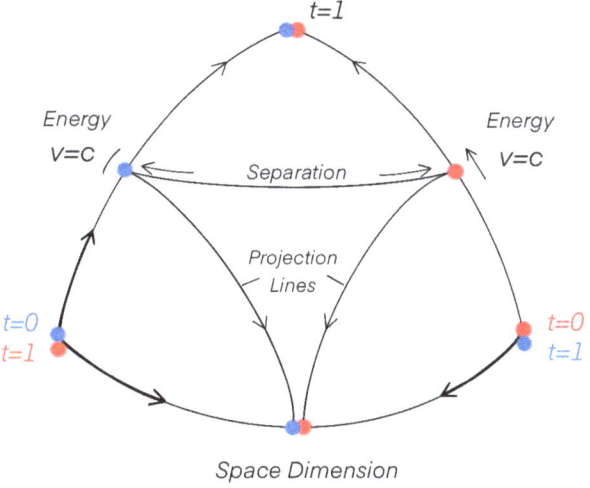

Figure 4.8. Light closing speed

You may need to think about this demonstration for a while . . . but it does explain how light always goes at the speed of light, as per the theory of relativity. For it to do this, there must be a fourth dimension orthogonal to the three of space, and light must travel through it as it travels through space.

The same result can be obtained in flat space with two cars travelling on separate highways that meet at a 60-degree angle. Both cars can have a velocity of 100 km/hr, but their closing speed, the rate that their absolute separation in space changes, is only the same 100 km/hr. This is a trick of geometry in flat space. The relativistic phenomenon is a trick of geometry in curved space . . . a four-dimensional Euclidean sphere, in fact; and the Lorentz transformation is equivalent to a rotation of this sphere.

Note also that two particles can be together in the energy dimension but appear separated in the three dimensions of space. This may shed light on quantum entanglement and what Einstein called "spooky action at a distance." We will explore this idea later in the book.

NO TIME

To reiterate, time is *not* the fourth dimension in this book. We killed time a few pages ago, or at least relegated it to the status of a ratio of distance to velocity. But time is a stubbornly persistent illusion. Distance can be measured in the three space dimensions, and velocity is a change in position in those three dimensions with respect to time. Have we somehow come full circle with the reintroduction of time?

No, because time can be removed from the picture by taking velocity as a change of position relative to other objects in motion. For example, we could say that a comet has moved a certain distance while the Earth completed one orbit of the Sun, without calling that one revolution a year. A motion-to-motion, or energy-to-energy, comparison bypasses time. The comparison should yield the same result for any observer. Welcome to the perspective of the third sister of the triplets we met earlier; she saw two events with different subjectively measured times taking the same time.

OTHER FOUR-DIMENSIONAL REPRESENTATIONS

Kinetic energy, or energy due to motion, in the realm of the low velocities we encounter in everyday life, is given by the formula,

$$E_k = \frac{1}{2}mv^2$$

Kinetic energy, in classical physics, is equal to half the mass times the velocity squared. For a given mass, the kinetic energy is dependent on (is a function of, in math terminology) velocity. In this sense, velocity is energy.

Hermann Minkowski (1864–1909), a Polish/German mathematician, was one of Einstein's professors. He demonstrated how the special theory of relativity could be expressed geometrically in terms of four-dimensional space-time. Minkowski space-time is a four-dimensional manifold, described by the equation,

$$ds^2 = -c^2 dt^2 + dx^2 + dy^2 + dz^2$$

In this equation, (*s*) is an arc length in space-time, and the time dimension is time converted to units of length by multiplying it by the speed of light (*c*). Time in seconds multiplied by metres per second yields a distance interval in metres, compatible with the intervals of distance in the three dimensions of space. This formula is an extension of the Pythagorean theorem from two to four dimensions.

Henri Poincaré (1854–1912) wrote a paper on relativity in 1906 that showed that if time were regarded as an imaginary term (an orthogonal axis to the real axes of space, designated by (*i*), then the Lorentz transformation becomes an ordinary rotation of the four-dimensional Euclidean sphere. This becomes rather complex, so let us just go with the explanation given so far in this chapter, but use the following equation for complex Minkowski space-time:

$$x^2 + y^2 + z^2 + (ict)^2 = const$$

We want to try to replace the time dimension with an energy dimension and get a space-energy equivalent to Minkowski's formulation for space-time. To do that, we need a term for the energy dimension having the unit of length, just as the x, y, z terms are distances in the three dimensions of space. There are two ways of achieving this goal.

The first way is by using wavelength (λ) as a proxy for energy and because it has dimensions of length. The wavelength of an electromagnetic wave is related directly to its energy according to the formula,

$$E = hf = hc/\lambda$$

where (*h*) is Planck's constant, (*f*) is the frequency, and (*c*) is the speed of light.

Thus, the term $(ict)^2$ in the complex Minkowski space-time equation can be substituted with $(i\lambda)^2$.

Note that the (*i*) denotes the imaginary part of a complex number, and (*i*) is equal to $\sqrt{-1}$. A discussion of complex numbers is available by searching the term online. A list of keywords appears in this book's Appendix of Keywords, for those who wish to become familiar with this area of mathematics. For now, please just trust that this (*i*), or $\sqrt{-1}$, represents an orthogonal fourth dimension.

The second way of getting an energy dimension term is by recognizing that,

$$Energy = Force \times Distance$$

And therefore

$$Distance = Energy/Force$$

Mathematically,

$$d = \frac{E}{F}$$

in metric units of ($kg * m^2 / s^2$) / ($kg * m^2 / s^2$) = m, metres.

The next chapter will explain exactly what forces are, but for now, we can just replace the (ict)^2 term with one that represents length in energy terms, $\left(\frac{iE}{F}\right)^2$.

This allows the creation of a new four-dimensional manifold in space-energy that may prove useful.

We can define distances in this space-energy as

$$x^2 + y^2 + z^2 + (i\lambda)^2 = constant$$

or

$$x^2 + y^2 z^2 {}_+ \left(\frac{iF}{E}\right)^2 = constant$$

Why go to the trouble of creating this new space, other than spaces are the final frontiers? As described in the last chapter, it is related to an entirely new way to look at gravity that does not require the curvature of space or time.

Of course, there is nothing wrong with time as the fourth dimension. The use of energy is just a different way of explaining the same observations. An alternate model or world view is simply adopted. However, when it comes to explaining forces such as gravity and the behaviour of light when it refracts as in a prism, the energy perspective may have significant advantages.

The space-energy version of Minkowski space-time also brings us a step closer to the theory of relativity, the ultimate description of the nature of the

Universe. The Minkowski formulation is completely consistent with relativity, and by extension, perhaps these new formulations are as well.

Later, we will see that the presence of a fourth dimension of energy opens the door to possible explanations of strange quantum mechanical occurrences such as entangled particles and "spooky action at a distance," as Einstein called these phenomena.

Next, we will investigate what forces really are.

FORCE

5

FORCE

Isaac Newton's second law of motion is given by the formula,

$$F = ma$$

where a force (F) acting on a mass (m) gives it an acceleration (a).

Acceleration is the derivative of velocity with respect to time (dv/dt), in calculus parlance. This is just a fancy way of saying acceleration is the rate of change of velocity. When you step on the aptly named accelerator in your car, its velocity or speed increases. Velocity is just a fancy way of saying speed that occurs in a definite direction. Velocity is a vector quantity, meaning it has a direction as well as magnitude, for instance 100 kilometres per hour due north. Speed is a scalar quantity, just a number, like 100 km/hr. Scalar is just a fancy way of saying . . . oh my, stop with the fancy ways!

What is important is that acceleration refers to a change in velocity with respect to time, and since we killed time in the first chapter, we need to state the force equation in a timeless format. To get at the meaning of force, we need a formulation that refers only to energy. Luckily, there is a means to do this:

$$F = ma = k \times \frac{dE}{dx}$$

or

$$F = k \times \frac{dE}{dx}$$

where force (F) is directly proportional (k) is a constant of proportionality to the derivative of energy (E) with respect to distance (x), or the change in energy with distance, the energy gradient in space.

The element of time has been removed. We will set about showing how this is so with some examples. Perhaps Obi-Wan should have said, "May the energy gradient slope away from you!" but it does not quite roll off the tongue as nicely.

Taking a step back from the math for a moment: What is the point of defining what a force is in this manner? In fact, it is just a different way of looking at things—a new paradigm—and maybe more useful for our purposes. Any force can be attributed to an energy gradient. A simple example might help to ease into the concept, one that will be familiar to Star Wars fans.

When someone is thrown back by the force of an explosion, it is because there is more energy concentrated in the explosive device than in its surroundings. When that energy is released, it seeks equilibrium with the energy in the surrounding environment. Eventually things cool down; the heat of the explosion dissipates into the environment. The temperature, or heat content, of the environment is related to the velocity of the molecules that make up the system. Temperature is the kinetic energy of molecules as seen from a large-scale perspective. Initially, there is quite a steep energy gradient away from the bomb that produces the explosive force of the blast.

Gravity, in classical physics, is another force, so we should expect that the force of gravity is equal to the derivative of the gravitational potential energy with respect to distance in space. This expectation is satisfied. In calculus, taking the derivative of a function is finding its slope or gradient at any point, and it reverses the process of integration discussed previously. Let us look at the formulas from the last chapter to clarify.

$$U = -\frac{GMm}{r}$$

Where (U) is the gravitational potential energy and (r) is the distance between two masses (M) and (m). As before, (G) is the universal gravitational constant. In these formulas, (r) and (x) are used interchangeably to represent radial distance and linear distance, respectively.

$$F = \frac{GMm}{r^2} = dU/dr$$

Where (F) is the force of gravity, a force that gets weaker with the square of the distance (r) separating the two masses. In calculus, (dU/dr) represents the derivative of energy with respect to distance, the energy gradient in space.

Below we see that integrating the force (F) with respect to distance (r) gives the potential energy change (U) over that distance.

$$U = \int_{R}^{\infty} \frac{GMm}{r^2} dr = -\frac{GMm}{r}\Big]_{R}^{\infty}$$

Capital (R) is the lower limit of integration, the initial radius we would use to integrate from in summing up work or energy required to move an object from that proximity out to infinity.

Similarly, by differentiating (U), we arrive back at the force (F).

Now, employing the new formula

$$F = k \times dE/dx$$

we should expect,

$$F = \frac{GMm}{r^2} = k \times dE/dr = k \times dU/dr = k \times \frac{GMm}{r^2}$$

to be true, and it is! For those without a calculus background, trust for now that this is correct.

All we are saying, besides give peace a chance, is that the force of gravity is due to the slope of the contours of the energy background, the fourth dimension.

Significantly, we have found a way to define a force without referring to time, bypassing the time related acceleration in Newton's second law—a step closer to defining gravity without curved space-time.

SPRING IS HERE!

Let us look at another example, familiar to students of physics: a spring.

Springs are governed by Hooke's law, which was developed by Robert Hooke (1635–1703):

$$F_s = -kx$$

where the force (F_s) required to compress a spring over certain distance is equal to the spring constant (k) times the distance (x) that the spring is compressed.

Suppose we send a mass hurtling toward the spring with kinetic energy of $E_k = \frac{1}{2}mv^2$. If we ignore the heat and friction loses, the spring will contain this exact quantity of energy when it has compressed enough to bring the mass to a dead stop, its velocity zero, its kinetic energy fully transferred into the spring. Figure 5.1 pictures this scenario.

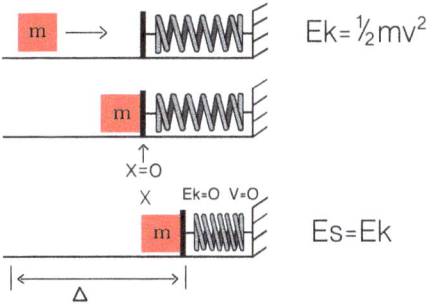

Figure 5.1.1 Spring and mass

Hooke's law and the fact that energy (same thing as work) is equal to force multiplied by distance, suggests that the energy in the spring is the integral of the force on the spring over the distance it is compressed.

The expression for that is

$$Work(W) = Energy(E) = \int_0^x -kx\ dx = \frac{1}{2}kx^2$$

Since the energy that is stored in the spring when it has stopped the mass is equal to the kinetic energy that the mass originally had, ignoring friction losses, we can say,

$$\Delta E / \Delta x = \frac{1}{2}mV^2/x$$

the change in kinetic energy over the distance the spring is compressed.

Since the kinetic energy is the transferred to the compressed spring during the displacement through distance (x), we can conclude that,

$$\frac{1}{2}k\frac{x^2}{x} = \frac{1}{2}m\frac{V^2}{x}$$

which reduces to,

$$k'x = \Delta E / \Delta x = Fs$$

The (k') term is just a different constant incorporating the factor 1/2. We are now back to Hooke's law and have shown again that the force in a spring is the energy gradient, the change in energy divided by the compression distance.

A VECTOR MAKES A POINT

A further example can be produced by looking at the change in the kinetic energy vector of a mass on a string being whirled in a circle. The vector changes by a small amount as the mass moves an infinitesimal distance around the circumference of the circle. The differential of these two energy vectors is a tiny vector pointing into the centre of the circle, which is the centripetal force. The diagram below illustrates this.

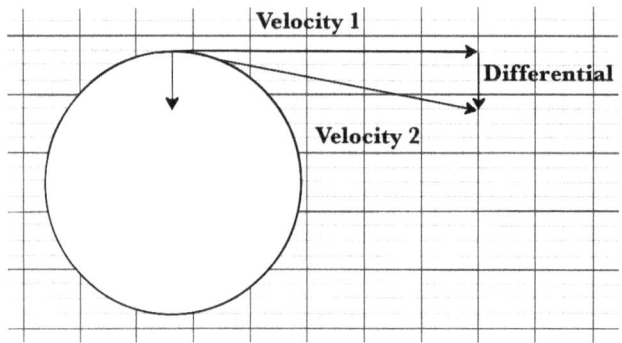

Figure 5.2. Mass in circular motion

ANTI-ANTIGRAVITY

The thrust of this chapter has been to demonstrate that all forces are energy gradients. Hopefully, reinforcement to the idea that gravity is just a gradient or slope in the energy background intensity. Next, we move on to a new topic: the speed of light. One final somewhat disappointing thought first, though. . .

Since we have discovered what gravity really is, we now know what antigravity must be: *the addition of energy to reverse the energy gradient.* If this was a book

FORCE

about anti-gravity you couldn't put it down . . . but, moving right along . . . When you apply energy to your coffee cup to lift it off the desk, you are using antigravity. When we strap a satellite to a rocket to give it the potential and kinetic energy to stay in orbit, that is antigravity. To counter gravity, we must provide sufficient energy to offset or reverse the energy gradient. A weightless object has sufficient energy to match the background level, so there is no gradient and hence no force of gravity.

Therefore, antigravity is simply nullifying gravity by applying energy, and that is as good as it gets science-fiction fans!

SPEED OF LIGHT

6

SPEED OF LIGHT

The idea to be put forward next is known as a variable speed of light (VSL) hypothesis. The speed of light (c) in a vacuum is normally considered to be a fundamental constant of nature. Nothing can travel faster than light because of the huge energy requirement to accelerate anything with mass to that speed. Photons, though, the particles that light is presumed to be made of, are massless, so they can go zipping off at the speed of light.

The speed of light does vary though, in different media, but is fastest in a vacuum. Light (in a vacuum) travels at an astonishing 299,792,458 metres per second (these days). However, in other transparent media that light can propagate through, such as air, water, or glass, to name a few, light slows down to a degree. The ratio of the speed in a material to the speed in a vacuum is a property of the material called the refractive index. The idea that light travels at different speeds in different media is not novel at all.

Propagating through this chapter is the concept that the speed of light in a vacuum has varied over the life of the Universe since the Big Bang and has, in fact, been much faster in the past. Obtaining evidence for this is, of course, essential to the scientific testing of the entire hypothesis.

COSMIC INFLATION

Variable light speed and the mechanism behind it to be proposed here should allow us to dispense with an arbitrary assumption made by Big Bang theorists called cosmological inflation. I will just call it cosmic inflation because it is not logical, just an assumption without an identifiable cause. Cosmic inflation is an unexplained extremely rapid expansion of the Universe presumed to have occurred around 10^{-33} seconds into its life. The early rapid expansion is postulated to explain the observed uniformity of the early Universe.

We have reached the point in the story to reveal another astonishing finding. You will need to be armed with the background thus far, wherein the Universe consists of a volume of space containing mass and energy. The Universe has a certain amount of mass and a certain amount of energy, and since mass-energy is conserved, the total quantity is constant. The volume of the Universe, however, is expanding. The hypothesis describes it as a sphere expanding at the speed of light. The energy density and mass density are naturally declining as this expansion happens.

A FINITE UNIVERSE

That the Universe has a finite volume is debatable; some would say it is infinite in extent. The solution to Obler's paradox—that the night sky is not uniformly ablaze with starlight from an infinite extent of starry sky, where every line of sight terminates on the surface of a star—is normally resolved by saying the Universe is not infinite in extent.

For fun, I now digress into a quick mathematical argument that the Universe is in fact finite. The case for such a major conclusion about the size of the Universe is deceptively simple. We, you the reader and I, are the indicators that the cosmos is finite, believe it or not! Infinite is awfully big, so big that anything divided into infinity is zero. Nevertheless, we exist and are not nothing, not zero. Therefore, the Universe is not infinite! Well, the math works anyway.

VARIABLE LIGHT SPEED

Moving on, recall again that density is the ratio of mass to volume. Water density is roughly 1 gram per cubic centimetre. Energy density can be considered as a ratio of energy to volume. The Universe has an average volumetric mass density and energy density.

Turning to speeds, let us now consider the speed of something other than light: the speed of sound. The speed of sound is given by the equation,

$$c = \sqrt{\frac{K_s}{\bar{\rho}}}$$

where (c) is the speed of sound from the Latin celeritas, meaning velocity. This (c) is not to be confused with the (c) representing the speed of light, for a moment anyway. The (K_s) stands for coefficient of stiffness, the isentropic bulk modulus, just an indicator of how much a material resists being deformed. Air has little resistance to being deformed; steel has a lot more resistance to having its shape changed. The term $\bar{\rho}$ stands for average mass density, the ratio of mass to volume.

Examining the formula, we can see that sound travels faster in a stiffer material and slows down in higher density material. To give some idea of the various speeds of sound, refer to the table in Figure 6.1. Observe from the data that the speed of sound in the same media, air, varies with the temperature/energy content of the air.

Material	Speed of Sound (metres/second)
Rubber	60
Air @ 0 °C	331
Air @ 20 °C	343
Air @ 40 °C	355
Gold	1200
Copper	2260
Aluminum	3100
Glass	4540

Figure 6.1. Speed of sound in materials

Sound moves very slowly through rubber, a supple material easily deformed yet high in density. On a cold day, sound moves more slowly through air than it does on a hot day, because the density of the air increases as it gets colder. The speed of sound in metals is relatively fast because metals are stiff, and as they get less dense, sound really moves along.

A DERIVATION AT LIGHT SPEED

What does this have to do with the speed of light?

Begin with the idea that the Universe has a mass density and an energy density. Using the formula for the speed of sound, let us substitute in these quantities:

$$c = \sqrt{\frac{K_s}{\rho_u}}$$

or

$$c = \sqrt{\frac{K_s}{mass_u / volume_u}}$$

where the subscript indicates that we are referring to the Universe as a whole.

The Universe also has an energy density. Why not assume that has something to do with its stiffness or bulk modulus? Assumption made; we substitute energy (E_u) / volume (v_u) for (K_s). This brings us to the following equation,

$$c = \sqrt{\frac{E_u/v_u}{m_u/v_u}}$$

Canceling the volumes, we get

$$c = \sqrt{\frac{E_u}{m_u}}$$

Eliminating the square root by squaring both sides, we obtain the following in general terms:

$$c^2 = \frac{E}{m}$$

or the much more familiar

$$E = mc^2$$

Have you ever seen a more concise derivation of that equation? The possibility that it concerns densities is not obvious because the volume terms have cancelled out and are no longer seen. Well, the math works anyway.

In that sense, it is a hidden variable theory, where volume is the hidden variable. The variance of the speed of light during the life of the Universe is now quantified and qualified in a manner consistent with the theory of relativity.

Furthermore, the suggestion is that light is propagating through a media, raising the historical debate about the aether. Yes, the aether is back, but it is energy itself, with the added twist that the energy dimension is orthogonal to the three dimensions of space. This will be shown to negate the results of experiments that supposedly proved there is no aether, such as the famous Michelson-Morley experiment (1887).

Figure 6.2 illustrates the proposed four-dimensional space-energy in graphical form. The x-axis represents any or all the three dimensions of space, and the y-axis represents the orthogonal energy dimension.

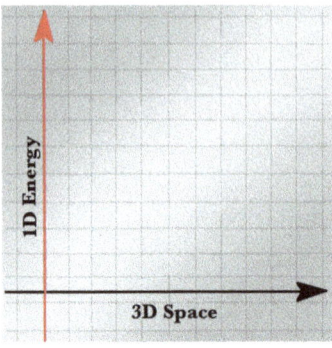

Figure 6.2. Space-energy

The idea proposed is that light propagates through a sea of electromagnetic energy, an energy soup if you will, with those curly noodles as electromagnetic waves. Like a strand of spaghetti wiggling through a whole pot of spaghetti, but the pasta is electromagnetic waves. Just a food analogy for thought.

Well, that is cool and all, but what does it mean? Remember from the theme song of the TV show *The Big Bang Theory* written and recorded by the Barenaked Ladies that "our whole Universe was in a hot, dense state"—the theory that the young Universe was all energy, compacted into a small space. Initially, it had an extremely high energy density and no mass at all. Mass came later as the Universe expanded and cooled. The implication for the speed of light is that initially,

$$c = \sqrt{\frac{high\ energy\ density}{near\ zero\ mass\ density}} = near\ infinite\ speed\ of\ light$$

As the Universe expanded and cooled, some mass formed and the energy density decreased, causing the speed of light to slow:

$$c = \sqrt{\frac{lower\ energy\ density}{non\ zero\ mass\ density}} = significantly\ lower\ speed\ of\ light$$

to the value we accept today,

$$c = \sqrt{\frac{curent\ energy\ density}{current\ mass\ density}} = 299,752,458\ m/sec$$

When propagating through denser media such as glass, light slows down relative to its speed through a vacuum. Normally, a perfect vacuum is space containing no atoms, no mass. There is a background energy known as zero-point energy or vacuum energy, but its magnitude varies greatly, and it may not be related to the background energy hypothesis of this book. The presence of electromagnetic radiation should not be conflated with virtual particles or quantum fluctuations.

As was pointed out earlier, space is full of electromagnetic energy, even if it doesn't meet the eye. And the true depth of the sea of energy is unknown; what we experience are the relative surface level, the peaks and troughs at the top.

TESTING, TESTING . . .

At the outset of this story, it was claimed that there would be tests to be carried out on these ideas in the spirit of true science. The foregoing leads clearly to one of these tests, but it requires an extremely precise measurement of the speed of light over a period of many years. If the speed of light in a vacuum can be determined to be measurably decreasing over a period in the lifetime of the Universe, it would be evidence in favour of the idea of variable speed of light produced by the mechanism explained above.

Keep in mind that the Universe has been around an estimated 13.8 billion years, or perhaps longer based on recent observations, while humans have been measuring the speed of light for only about 400 years. Galileo Galilei (1564–1642) tried first. The Danish physicist Ole Rømer (1644–1710) determined a reasonable approximation in 1676. Albert A. Michelson conducted measurements up until his death in 1931. Even 400 years is only 0.000003% of the life of the Universe, and there has always been experimental error and uncertainty in these measurements.

Have we been measuring the speed of light with enough precision and for long enough to notice any tiny changes in its speed? No, we have not, in the author's opinion. Therein lies the first proposed test of this theory to determine if the speed of light in a vacuum has been slowing over the history of the Universe.

Light speed would have decreased extremely rapidly in the very early Universe, when energy density was rapidly declining due to expansion happening explosively. The speed would have declined dramatically, then begun to level out. The volume of a spherical Universe is proportional to the cube of its radius, and the radius would have been increasing at the speed of light, according to the proposal of this book. The year-over-year change in the speed of light at the present time might be very small indeed. A rough estimate of the current rate of change is 0.00015 metres/second/year.

We are certain that light travels slower in the presence of mass: slower in glass than in air; slower in air than in a vacuum. There is no question that the speed of light is variable.

SPEEDING THROUGH THE LIFE OF THE UNIVERSE

Let us look at a diagram in Figure 6.3, portraying the life of the Universe.

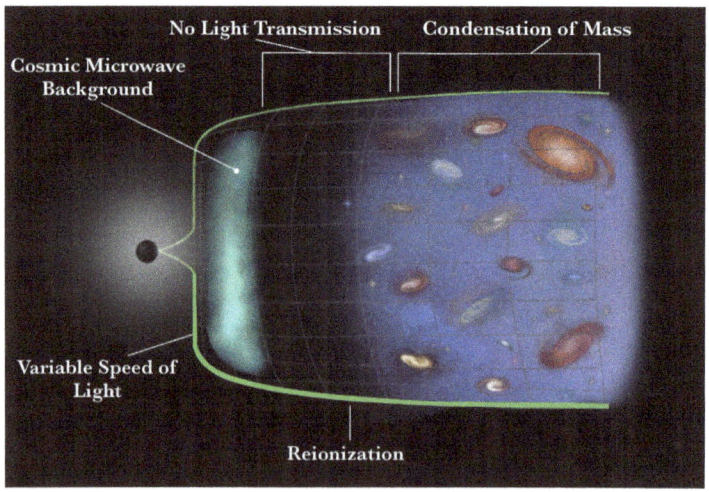

Figure 6.3. Life of the Universe

One simple observation is that the green outline of the bottom of this bell shape that looks a bit like a megaphone approximates the shape of the graph of the speed of light that one derives by assuming it varies with the energy density within a sphere expanding at the speed of light. The graph below shows how that would look.

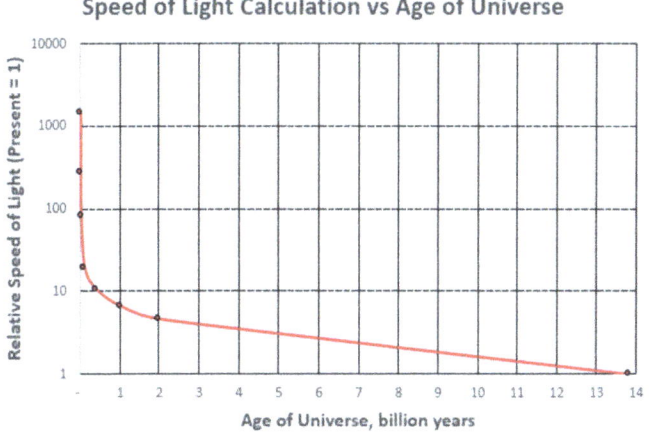

Figure 6.4. Speed of light vs age of Universe

The Big Bang can be roughly visualized like the explosion of a fireworks shell in the night sky. Everything rapidly expands, then slows and cools. Of course, with fireworks, the slowing of the expansion is caused by air resistance, and this is not the case with the Universe. The author hypothesizes that the cosmos has always expanded at the prevailing speed of light, slowing contemporaneously with the reduction in the energy density caused by the expanding volume.

COSMIC INFLATION

The Universe is remarkably uniform (homogenous and isotropic) in every direction, as seen in the cosmic microwave background (CMB) radiation, the ancient energy signature of the Big Bang. There is thermal equilibrium in all directions, but there is no time for this to occur. Distant regions in space are so far separated by standard Big Bang expansion that they could never have been in causal contact because standard light speed travel time exceeds the age of the Universe. This quandary is known as the horizon problem.

Theorists therefore postulate that the Universe expanded suddenly and dramatically shortly after the Big Bang from a tiny uniform structure to a huge uniform structure at a rate of expansion much faster than the speed of light is today. The expansion due to cosmic inflation is thought to be factor of 10^{26} and to have occurred between 10^{-33} and 10^{-32} seconds after the Big Bang. For perspective, this would be like one nanometre suddenly becoming 10.57 light years: an expansion much faster than that of the national debt!

If instead the speed of light itself was much, much faster early in the life of the Universe, would that not just as easily explain its uniformity today? The question is a fair one since inflation is just a convenient assumption with no known cause.

An advantage to the VSL concept is that it eliminates the need for cosmic inflation. Essentially, the higher speed of light early in the life of the Universe could achieve the same effect, a very rapid expansion such that everything is close to being uniform.

Indeed, many variable speed of light theories have been proposed, and a summary of them appears in the article "New Varying Speed of Light Theories" (2003), by João Magueijo. Many complexities, mathematical and philosophical, come with these theories, so a detailed review of them here is beyond the breadth of this book.

The VSL hypothesis proposed by the author at least passes the test of Occam's razor in that it is relatively straightforward, fitting neatly into the book you are holding. Further tests to prove or disprove the idea will also be proposed later in a subsequent chapter.

7

AETHER

"Only the existence of a field of force can account for the motions of the bodies as observed, and its assumption dispenses with space curvature. All literature on this subject is futile and destined to oblivion. So are all attempts to explain the workings of the Universe without recognizing the existence of the aether and the indispensable function it plays in the phenomena."

— Nikola Tesla (1856–1943)

AETHER

Well said Mr. Tesla! Aether it exists, or it does not. Among the contentions in this book is that aether is in fact alive and well. This may not sit well with mainstream scientists because the aether was done away with by the Michelson-Morley experiment, but this version is unique.

The original aether was envisioned as the medium to carry light, just as sound travels through the air. Light is a wave; it should not be able to travel through a vacuum, they thought, but it clearly does. There should be something waving, just as waves on the ocean require water to give them form. The existence of the aether was vigorously debated up until the famous Michelson-Morley experiment in 1887 apparently demonstrated it did not exist. More will be said about this in the chapter that follows this one.

In Chapter 6, "Speed of Light," we derived ($E = mc^2$) from an equation that was developed for sound waves travelling through various media. A derivation of Einstein's equation relating energy to mass through the speed of light was accomplished by assuming the media is characterized by its energy density and its mass density. A fourth dimension of energy, orthogonal to the three dimensions of space, was earlier conjectured.

Now the task is to demonstrate, using a thought experiment, how an orthogonal energy dimension is virtually essential to validate the relativistic prediction that light always moves at the same speed regardless of how the observer is moving.

Please refer to Figure 7.1, containing a diagram of the setting for our thought experiment. Imagine, if you will, a point A and a point B, separated by one light second in distance. This is about 80% of the distance from the Earth to the Moon, for scale. One light second is the distance light moves in one standard second. Imagine, then, a stationary observer located at the vertex of an equilateral triangle, labelled point O, for "observer."

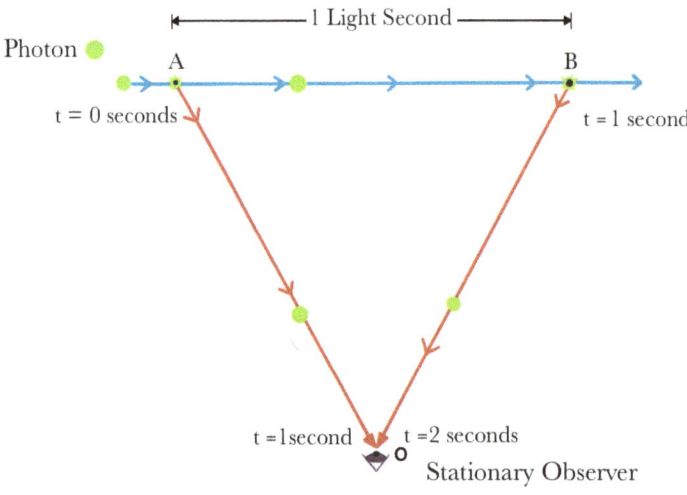

Figure 7.1. Stationary observer case

A photon of light appears from the left and comes down the blue line from A to B. When it encounters point A, it triggers the emission of another photon directed from A to the observer, located one light second away at point O. In the lab, a beam splitter could be used to implement the diversion. Start the clock at ($t = 0$), when the original photon arrives at A.

The original photon then continues to B, reaching that location one second later and triggers the emission of another photon aimed directly at observer O. The time of this event will be one second after the photon hit A, so now time is ($t = 1$). Simultaneously, the secondary emitted photon from A reaches observer O. One second later, the secondary photon from B reaches observer O, at ($t = 2$). The observer sees a flash from A, then a flash from B one second later, and is thus able to ascertain that A and B are one light second apart, or 299,792,458 metres distant.

Next, consider what happens if the observer is moving at with an arbitrary velocity along a line perpendicular to line A-B, within the plane defined by the three points A, B, and O, as shown in Figure 7.2. Everything else in the experiment remains the same as previously, other than the motion of the observer. All the photons stay within the plane ABO, the sheet of paper that the drawing is on.

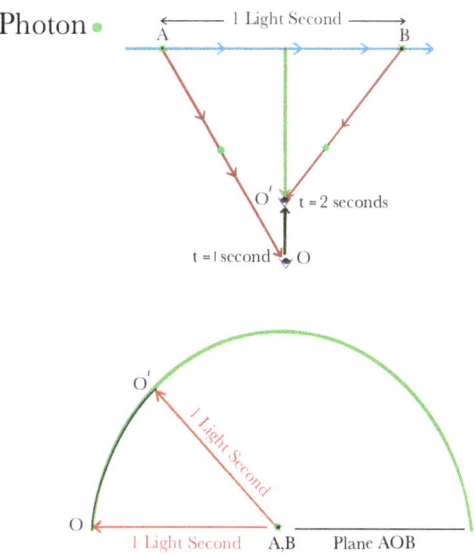

Figure 7.2. Moving observer case (top view and end view)

The original photon hits point A at ($t = 0$), triggering a secondary photon that intercepts the now moving observer precisely at point O by design, and time ($t = 1$) second. As before, the original photon arrives at point B and sends a photon off to where O has moved to in the interim, point O'.

A problem arises, because the distance is now shorter from B to O' than from B to O, and the photon from B arrives at the observer's new location in less than one second. The observer knows that point A and point B are one light second apart. He would have to conclude that the speed of light is not the same while in motion as it was when stationary, having seen the two flashes less than one second apart in this version of the experiment.

Our observer has done the experiment from a stationary position at point O and knows the size of the experimental layout; knows, too, that the speed of light is a constant. Something must be wrong, because according to relativity, all observers regardless of their motion should find the speed of light to be the same. Relativity has been tested and proven true; it is inconceivable that relativity is wrong.

Fortunately, there is a way out of this conundrum, provided there is a fourth dimension of energy, orthogonal to the three dimensions of space. When an observer is in motion, they are in effect displaced in the energy dimension, relative to the equipment in the experiment.

Returning to the experimental layout, we see that points A, B, and O are a triangle that defines a two-dimensional plane in three-dimensional space. Now suppose point O' is not in that plane, but in a dimension orthogonal (90°) to that plane. The energy dimension in this experiment is perpendicular to the ABO plane of the paper.

Take a line from B to O' that is the same length, one light second, as the distance from A to B. If point O' is indeed out of the plane, that line can end on point O', one second later. And now everything goes round with respect to relativity. The observer sees the flash from A when passing through at point O, then sees the flash from B one second later at point O', and correctly concludes that A and B must be one light second apart. However, this only works when there is a fourth dimension, orthogonal to the three dimensions of space. Admission to that dimension is gained by possessing energy, in the form of a velocity, with respect to the experimental layout.

In fact, there is a circle that photons from A and B can reach each other in one light second. That circle is in a plane that bisects the line A-B and is perpendicular to the line and to the plane of the paper. The end on view of line A-B in figure 7.2 shows part of the circle.

So, the aether, in a sense, is back. The revived aether is a very strange media indeed. In fact, it is energy itself, the soup of electromagnetic radiation that we exist in. In addition, it is a dimension orthogonal to the normal three of space, at least mathematically. No wonder efforts to detect it were unsuccessful!

Could we say the observer is experiencing dilated time as the fourth dimension? Perhaps time has slowed because the observer is moving? Yes, that is the relativistic interpretation. However, we can cast ourselves in the role of the triplet sister on the Moon. We can look down on the experiment and see that the speed of light measurement should differ for the stationary and the moving observer by virtue of the unequal light path lengths B-O and B-O'. We know that time is relative. In the space-energy domain, the light path lengths are the same and the measurement of light speed is constant.

TIME, AGAIN

So, what has really changed? Is it the observer's time, or is it the observer's energy? Does the granular sugar dissolve faster due to more surface area or because the energy to crush the sugar cube has already been put in, as was the question in the story in the introduction to this book?

Present-day conventional wisdom will choose the time dilation explanation. Despite that, the energy explanation is a better fit for explaining gravity, or at least it is considering the arguments that all forces are energy gradients and that the energy locked up in mass is exactly that required to eliminate the potential energy of its gravity well.

The time explanation involves curved space-time; the energy explanation is straightforward in all four dimensions. These are two different ways of looking at the same thing.

Just as the Lorentz transformation accounts for time dilation, it similarly accounts for mass increase as kinetic energy is added to an object. As we discussed earlier, it is mathematically equivalent to rotation into an orthogonal dimension and geometrically equivalent to the Pythagorean theorem.

We have, within the new model, explained gravity as a force due to an energy gradient. The gradient is a slope variation in the new aether.

WHY?

Is making this distinction worthwhile? Why not just stick to curved space-time? Only because there seem to be significant advantages to the space-energy perspective and the VSL hypothesis that accompanies it.

Among the advantages yet to be shown are the following:

- an explanation of inertia and how it operates.

- an explanation of entropy and reasons for its general tendency to increase.

- elimination of three major mysteries: cosmic inflation, dark matter, and dark energy.

- a direct tie to gravitational waves.

- elimination of singularities in black holes and a reason light really cannot escape them.

- a prediction about the fate of the Universe.

These seem sufficient reasons to continue. The next section will discuss how the Michelson-Morley experiment failed to detect any aether.

NO AETHER

8

"The aether: Invented by Isaac Newton, reinvented by James Clerk Maxwell. This is the stuff that fills up the empty space of the Universe. Discredited and discarded by Einstein, the aether is now making a Nixonian comeback. It's really the vacuum, but burdened by theoretical, ghostly particles."

— Leon M. Lederman (1922–2018)

"No, it isn't!

— The Black Knight, Monty Python and the Holy Grail (1975)

"The medium is the message."

— Marshall McLuhan (1911–1980)

NO AETHER—THE MICHELSON-MORLEY EXPERIMENT

The aether was envisioned as the medium through which light propagates, just as sound travels through the air. Light, being a wave, should not be able to travel through a vacuum, but it does. Every day experience suggested there must be a carrier substance, just as waves on the ocean require water to give them form. The existence of the aether was vigorously debated up until the famous Michelson-Morley experiment in 1887 indicated it did not exist.

The logic was that if light was moving through a medium (the aether) and the Earth was as well, then a difference in travel time should be observed if light was sent off in the direction of Earth's motion or at right angles to it. The configuration of the experiment to search for the aether was ingenious and would have worked in classical physics.

The experimental setup has a light beam sent from a source, then split to travel down two arms of equal length at ninety degrees orthogonal to each other. The two beams reflect off mirrors at the ends of the arms and arrive back at a point of comparison. The geometry of the situation is pictured in Figure 8.1.

The tell would be wave interference. If there was a medium that light was travelling through, the time taken, and distance light travelled, would differ between arms. The waves would arrive back slightly out of phase.

Careful study of the diagram shows that time taken to cover each path will differ slightly if there is a medium present. Interference is the pattern of partial reinforcement and partial cancellation that happens when waves are out of phase.

No interference and hence no difference in the speed of light was found, regardless of the direction light was sent relative to the motion of the Earth. The logical conclusion was that the aether does not exist. Einstein put the last nail in the coffin of the aether in 1905 with his special theory of relativity predicated on the speed of light being the same for all observers.

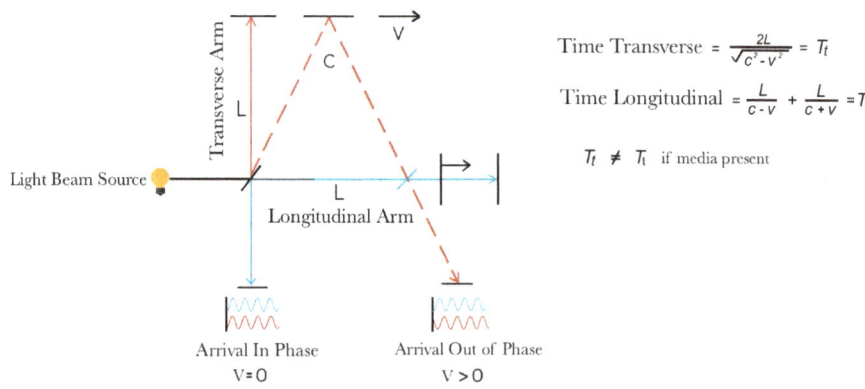

$$\text{Time Transverse} = \frac{2L}{\sqrt{c^2 - v^2}} = T_t$$

$$\text{Time Longitudinal} = \frac{L}{c - v} + \frac{L}{c + v} = T_l$$

$T_t \neq T_l$ if media present

Figure 8.1. The Michelson-Morley experiment

The contention here is that the Michelson-Morley experiment was misinterpreted. The results it showed were correct, but the reason is not that there is no aether. The reason is because the distance that light travels down each of the transverse arms of the apparatus in four-dimensional space-energy remains the same, regardless of the motion of the experimental device. This is a consequence of the Lorentz transformation, which cancels out exactly the difference in path lengths shown in the diagram.

Because of the motion of the Earth, there is a slight rotation into the energy dimension in the direction that it and the apparatus are moving. You will recall the energy axis *(v/c)* that arises when the Lorentz factor is represented as a right triangle that obeys the Pythagorean theorem. The total length (in four dimensions) remains the same for each arm, so light travels that same distance down each arm, comes back in synchronization, and shows no interference.

Mathematically, this can be proven. The equations from Figure 8.1 will be used to demonstrate.

The times for light to travel the transverse and longitudinal arms are given by,

$$T_t = \frac{2L}{\sqrt{c^2 - v^2}}$$

$$T_l = \frac{L}{c - v} + \frac{L}{c + v}$$

where (T_t) is the time for light to pass up and down the transverse arm, and (T_l) is time for light to move up and down the longitudinal arm. (L) is the length of the arm, (c) is the speed of light, and (v) is the velocity of the apparatus through the media.

The motion is along the line of the longitudinal arm. To make the times (T_l) and (T_t) equal, we multiply (T_l) by the Lorentz factor that specifies the rotation into the energy dimension, or length contraction in the space dimension.

$$T_l = \left(\frac{L}{c - v} + \frac{L}{c + v} \right) \times \sqrt{1 - \frac{v^2}{c^2}} = T_t$$

The easiest way to show this works is to plug in various numbers in a spreadsheet. It proves correct.

For example, let's use, for ease of calculation, $(L=1)$, $(c=1)$, and $(v=0.5)$.

$$\left(\frac{1}{0.5} + \frac{1}{1.5}\right) \times \sqrt{1 - \frac{0.5^2}{1^2}} = 2 \times 1/\sqrt{1 - 0.25}$$

$$\left(\frac{1}{0.5} + \frac{1}{1.5}\right) \times \sqrt{0.75} = 2 \times 1/\sqrt{0.75}$$

$$2.3094 = 2.3094$$

We saw in the last chapter that for the speed of light to be equal for all observers, stationary or moving, there must be a position that a moving observer occupies in energy as well as in space. If we limited the experiment to just the three dimensions of space, we got different speeds of light for different observers.

The obvious distinction in the case of the stationary observer and the moving observer is the velocity of the moving observer with respect to the experimental apparatus. This change is a discrepancy in the momentum and in the kinetic energy of the two observers. The Michelson-Morley experiment is the equivalent, except that it is the equipment that is moving along with the Earth and the light paths are equal length in four-dimensional space-energy.

To emphasize the point that the length stays the same in four-dimensional space even if the fourth dimension is time, we know that a moving object only appears to undergo length contraction as seen by a stationary observer. To an observer moving with the object, there is no change. All frames of reference are equally valid.

You might still be confused, wondering, is there an aether or not? It turns out that in the context of relativity, the aether is not necessary. However, under the space-energy paradigm, it is necessary. The aether is the energy background, and it has a density that governs the speed of light.

Where it gets strange is that electromagnetic waves compose both the medium and the light moving through it. Hence the Marshall McLuhan quote at the beginning of this chapter: "The medium is the message."

DARK MATTER

9

"You're entitled to say, if you're so smart, why don't you tell me what that dark matter is? And I'll have to confess I don't know."

—Jim Peebles (1935–)

COSMOLOGY—DARK MATTER

Galaxies, like our own Milky Way, are huge collections of stars that rotate like giant carousels in space. Astronomers have measured the velocity of the stars as they rotate around the centre of galaxies far, far away and have found something unexpected and problematic. The orbital speeds of stars when plotted against the radial distance from the centre of the galaxy produce what is known as a rotation curve or velocity curve. These curves often show a different velocity profile than can be explained conventionally with Newton's law of gravity.

The curves indicate constant or sometimes faster velocities with increasing radial distance from the centre of the galaxies, unexpectedly different from what the law of gravity dictates. Figure 9.1 shows an example, comparing an observed rotation curve to the curve the law of gravity predicts.

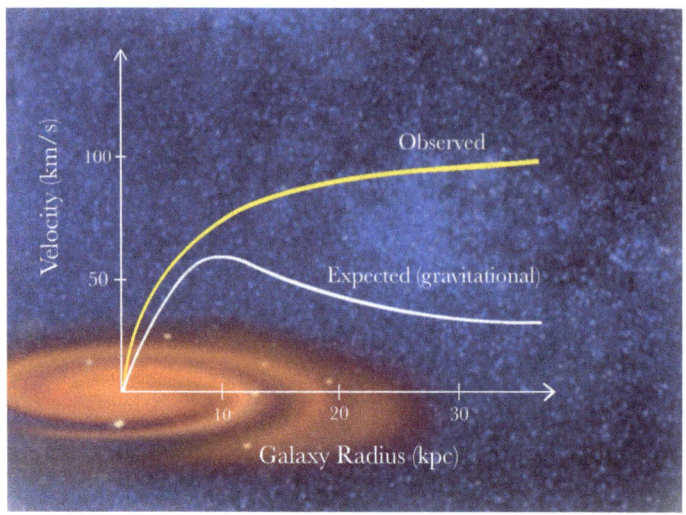

Figure 9.1 Galactic rotation curve

The problem is a massive one, since at the observed velocities the stars should not stay in orbit. Galaxies should fly apart because the stars are moving faster than they would if tethered by the force of gravity, based on the amount of visible mass estimated to be present. However, galaxies have persisted for billions of years, so something is amiss.

For a detailed discussion of galactic rotation curves, there is an excellent article titled "Rotation Curve Decomposition for Size-Mass Relations of Bulge, Disk, and Dark Halo in Spiral Galaxies" (2015), by Yoshiaki Sofue of the Institute of Astronomy at the University of Tokyo.

VELOCITY RELATIONSHIP—PLANETARY SYSTEMS

With a planetary system such as our solar system, where the mass is largely concentrated in the Sun and several large discrete masses we call planets, the orbital velocity is roughly proportional to the inverse square root of the radius of the orbit.

As shown below, the orbital velocity (v_o) of a planet is proportional to the square root of the universal gravitational constant (G) multiplied by the mass

(*M*) that the planet is orbiting, the Sun in our system, and inversely proportional to the radius:

$$v_o \approx \sqrt{\frac{GM}{r}}$$

For our solar system, the planet Mercury, closest to the Sun, moves at an average orbital speed of roughly 47.9 km/s, the Earth is slower at 29.8 km/s, and the furthest planet Neptune moves at a leisurely 5.4 km/s. The expectation, when astronomers began to look at galaxies, was that the same relationship for the orbital velocities of stars around galactic centres would hold true, but it does not.

DARK MATTER

The solution to this problem in vogue these days is to assume the existence of extra mass in the galaxies, several times more than can be accounted for by the matter that emits or reflects light. The visible stars, dust, and gases that are the known constituents of galaxies are all made of the ordinary matter familiar to us. The extra mass cannot be seen and so gets the name "dark matter."

Dark matter is thought to be about 85% of the total mass component of the Universe. The overall component split is 27% dark matter, 69% dark energy, and only about 4% is normal matter, or at least that is where the assumptions of dark matter and dark energy take us.

Dark matter has the amazing property that it is only detectable by its gravitational effects. There is no interaction between dark matter and normal matter or radiation, and it has never been directly observed. The addition of dark matter provides the extra gravitational attraction needed to hold the rapidly rotating galaxies together. However, the nature of this unknown substance is one of the greatest unsolved problems in present day physics and cosmology.

No one knows what the dark matter is, although various possibilities have been suggested. The suspects so far are axions, neutrinos, and weakly interacting massive particles (WIMPs) and gravitons. These are all hypothetical culprits and, in some cases, hypothetical particles. A lot of descriptive material is out there on these candidates, but we will not see any of it here because it is all dark. That was just dark humour. The author conjectures that dark matter does not exist!

Like the premise of cosmic inflation discussed earlier, dark matter is a means to explain the observations that only leaves us a with a bigger unsolved problem. Might there be a simpler explanation for the observed galactic rotation curves and the behaviour of the Bullet Cluster? The Bullet Cluster is a collision between two clusters of galaxies touted as the best evidence for dark matter. A few options for alternative explanations are now presented.

BETWEEN DISCRETE AND SOLID MASSES

Consider a planetary system, such as our solar system, where the masses in motion are discrete bodies, not connected except by gravitational bonds. The rotation curve describing the orbital velocity decreases as the inverse of the square root of the orbital radius, according to classical gravitational theory.

Next, consider a solid disk of material such as the base of a merry-go-round, one solid spinning disk of mass, with its individual atoms chemically bonded together. The velocity in this case increases linearly in proportion to the radius from the centre.

The rotation of galaxies seems to be an intermediate case, where rotational velocity is roughly constant with distance from the rotational centre. Does this suggest that galaxies are not quite made up of discrete masses in empty space, nor solid disks of material, but something intermediate, maybe discrete masses embedded in a liquid of some sort? Could the energy background proposed in this book have the characteristics of a liquid?

The author filled a bucket with water, spraying the water in along the inside edge of the bucket so that the water was left rotating slowly. Then a liquid soap was added that turned slightly grey, and the pattern it made looked like a spiral galaxy. Interesting but not really proving anything besides the floor needed mopping. The bucket was still present to keep things together.

A laboratory experiment would be necessary to determine and quantify the velocity profile of a liquid spinning in a container. The relationship might be revealing if it turned out to be a flat velocity curve as seen in the galaxies, because it would suggest that the proposed energy background is in some sense like a liquid.

THE MASS DISTRIBUTION SOLUTION

Consider the issue of the constant orbital velocity of the stars within a galaxy, using Newton's law of gravitation with a slight twist to allow for a particular galactic mass distribution. A mathematical excursion is needed to demonstrate this approach.

For a star orbiting within a galaxy, a force balance can be shown with an equation. The centripetal gravitational force that keeps a star in orbit must equal the centrifugal force that would otherwise cause it to fly off on a tangent to the orbit. Force is equal to mass times acceleration, and for circular motion, the radial acceleration is equal to the velocity squared divided by the radius.

We set the force of gravity by Newton's law equal to the centrifugal force on the star in orbit.

$$F_g = \frac{GMm}{r^2} = ma = m\frac{v_o{}^2}{r}$$

We can then solve for the orbital velocity and get the equation shown previously:

$$v_o \approx \sqrt{\frac{GM}{r}}$$

But what we observe is that the orbital velocities of stars are relatively constant with radius. So, we now force that to happen by saying,

$$v_o \approx \sqrt{\frac{GM}{r}} = Constant, K$$

This can be true if mass (M), the mass of the galaxy, is a function of its radius.

Think about the mass of the Earth as a function of the radius from its centre. The volume of a sphere, approximating the shape of the Earth, is proportional to the cube of its radius. Assuming Earth has a constant density, the mass is also proportional to the cube of the radius.

Returning to the galaxy problem, is there a function of mass (M), inside a sphere centred on the galactic centre, that allows the orbital velocity to be a constant with radius? There is, in fact, such a relationship. Mathematically speaking, the mass is a function of the radius:

$$M = f(r)$$

Specifically, the mass can be assumed to be a linear function of the galactic radius:

$$M = k'r$$

where (k') is a constant. This choice turns out to be appropriate.

Going back to the orbital velocity equation, we insert the mass function,

$$v = \sqrt{\frac{G(k'r)}{r}}$$

and we cancel the (r) term to obtain,

$$V = \sqrt{Gk'} = constant$$

This provides the answer we were looking for.

We can then find the density distribution within the galaxy if the mass is proportional to the radius,

$$M = k'r = \rho \times \left(\frac{4}{3}\pi r^3\right)$$

Where p is the density, and we solve for it, combining all constants into a new (k'') constant,

$$\rho = \frac{k''}{r^2}$$

This means that the mass density is proportional to (*1/r²*).

It should be noted that other density relationships could produce orbital velocity curves that are proportional to $\sqrt{r}$, for example.

The problem of the flat rotation velocities vanishes with a particular distribution of the mass such that the density decreases with the square of the radius. Perhaps the mass within galaxies is not distributed this way, but it is a nice simple solution.

I am not sure that we know exactly how the mass is distributed within galaxies, because galaxies presumably have massive black holes at their centres, a roughly spherical halo of gas and stars, orbiting clusters of stars, lesser black holes distributed throughout, less visible stars like brown dwarves, clouds of dust and gas, and other unseen objects made of normal matter. Mass distribution is assumed to be correlated to the luminosity profile of galaxies, but whether that gives a precise answer or not is open to question.

The galaxy constituents mentioned above are made of ordinary matter but may be difficult to see. Black holes can be dark because light cannot escape from them. Other objects may not emit sufficient visible light to be observable from our vantage point.

Surely, though, such a simple solution has been investigated and found wanting? I do not know.

INERTIA

Now is a perhaps a good time to introduce the concept of inertia. Any physical object has a resistance to a change in its velocity and hence its kinetic energy. Newton's first law of motion describes inertia as follows: an object will continue moving at a constant velocity unless it is acted upon by a force. Inertia is intimately related to the conservation of energy. Often it is described as simply a property of mass, like gravitational attraction.

The space-energy model provides a mechanism for inertia. Recall that gravity, in the proposed model, is a result of mass causing a dimple, a gravity well, in the energy background. When a mass is started in motion, or its velocity is changed, the gravitational well that accompanies it would have to be affected correspondingly. Inertia is, in this model, the resistance to that change by the media that fills the Universe.

Returning to our glass of ice-water Universe: If we want to move an ice cube, we must also relocate the accompanying depression in the water. The water will accommodate by moving out of the way and redistributing itself, but it takes a force to initiate the movement of the ice cube. That is the model equivalent to inertia in the Universe.

Gravity is a force that results from a gradient or slope in the background energy, and inertia is an accompanying resistance to changing the whole mass-energy landscape of the Universe. Gravitational mass and inertial mass are the same since they both are intimately linked to the presence of the gravitational well.

Consider now an approach to the problem of dark matter that incorporates a fourth dimension of energy. The electromagnetic energy background is the media that light travels through, and it has the properties of mass-energy density. The key point is that in this model there is a media.

The presence of inertia suggests that the energy background has a property akin to viscosity in a liquid. This brings us back around to the observation that rotation curves of galaxies seem to be a middle-ground case between discrete particles in orbit and solid disks spinning. Do galaxies exhibit rotational characteristics comparable to a liquid?

In addition, the energy background itself would have an equivalent mass, distributed in the shape of the gravity well of the galaxy.

These are the implications of a fourth dimension of energy, an ocean of energy we are all immersed in, if it truly exists. The suggestion is that gravity might behave differently on a galactic scale than on the scale of the solar system, because of the overall distribution of mass and interaction between gravitational wells and the energy background. We know that objects on the human scale behave much differently than the individual atoms that comprise them.

GRAVITY INSIDE A SOLID BODY

One more kick at the dark black cat. Gravitational potential energy decreases proportionately to the reciprocal of the radius *(U~1/r)* when we are outside the boundary of the object such as the Earth. However, if we go inside the object, the gravitational force behaves differently.

What happens if we go inside the body of any mass, such as that of our planet? The gravitational force becomes a linear function of the radius, with zero gravitational force at the centre. At the centre of the Earth, the mass surrounds us in every direction and its gravitational force is pulling away from the centre symmetrically, resulting in no net force.

Recall earlier we calculated the orbital velocity of a satellite above the Earth with the equation,

$$v_o \approx \sqrt{\frac{GM}{r}}$$

Imagine now if we could orbit inside the Earth, something possible only in a thought experiment. We know that gravity is no longer inversely proportional to the inverse radial distance squared from the surface of the Earth outward but is now proportional to the radius from the centre to the surface. The proof is as follows:

$$F_g = \frac{GMm}{r^2}$$

above the surface, a relationship with force proportional to *(1/r²)*.

Below the surface, if we substitute the spherical volume formula and a density *(ρ)* into the equation,

$$F_g = \frac{G\frac{4}{3}\pi\rho r^3 m}{r^2} = G\frac{4}{3}\pi\rho r m$$

the relationship is a force proportional to *(r)*,

And now the orbital velocity equation applicable inside the Earth becomes,

$$(v_o)^2 = G\frac{4}{3}\pi\rho r$$

or

$$v_o = k \times r$$

where the orbital velocity is proportional to the radius of the orbit and (k) is the constant of proportionality.

Note that the orbital velocity inside a mass—were it possible to orbit inside a mass—has the same relationship to radius as the velocity of a point on a rotating solid object.

Suppose we consider the stars in a galaxy to be "inside" the galaxy that contains them. Would that mean that their orbital velocity should be proportional to their distance from the centre and that light curves should not be flat but show increasing velocities with radius?

Galaxies are not solid masses, but they are concentrations of mass relative to the emptiness of intergalactic space.

Here is an interesting thing to contemplate. If we were to expand the solid mass of the Earth to the size of the Milky Way galaxy, an expansion of roughly 3.6×10^{13} times in diameter, the Earth's consistent atoms and molecules would be distributed in mostly empty space. Atomic nuclei would be roughly 0.036 metres in diameter and separated by roughly 3600 metres, a ratio of 100,000 times of their separation to their diameter.

Taking the average separation of stars in the Milky Way as 5 light years or 4.7×10^{13} metres, and the diameter of an average star of 1.4×10^9 metres, we get a ratio of 33,000 times of separation to diameter.

The scaled-up Earth is comparable to a galaxy in terms of the ratios of separation to diameter of their constituent atoms or stars respectively. There is a factor of three difference, but still remarkably similar considering the comparison is cosmic scale to atomic scale. In this sense, a galaxy of stars is comparable to a planet of atoms.

SUMMARY—GALACTIC ROTATION CURVES

We are left with the observed light curves showing roughly constant star velocities with increasing radius ($v_o = constant$), an intermediate case between discrete masses in orbit ($v_o \sim 1/\sqrt{r}$ *relationship*), and a spinning solid object or an orbiting object inside a mass ($v_o \sim r$ *relationship*). We have the observations surrounded!

Perhaps the most defensible means of explaining the uniform orbital speed of stars in galaxies is the mathematical result with a mass distribution such that the mass of the galaxy is linearly proportional to its radius and the density decreases with ($1/r^2$). The result is a constant orbital velocity as a function of radius, matching the observed rotation curves.

All that is required to explain the observations, without invoking dark matter, would be the appropriate galactic mass distribution, or even one that approximated it, since the light curves do have some non-linearity.

The belief is that most galaxies have black holes at their centres, but how prevalent they are out among the surrounding stars is still unknown. How massive are these black holes, and what do they do to the overall mass distribution profile? The largest known black hole is a quasar (quasi-stellar object) TON618 that may contain over 40 billion solar masses. Even larger black holes are possible.

How much unseen regular matter is distributed within galaxies? For example, Jupiter-like giant planets that do not quite have enough mass to ignite the nuclear reactions needed to emit visible light. How much dust and debris are out there, and how is it distributed? And, of course, light itself, the energy background, has an equivalent mass. Are we taking that into consideration?

Is it easier to believe that the mass in galaxies tends to be distributed according to the relationship that works to match the observations, or that 85% of all the mass in the Universe is something that we cannot detect, that does not interact with normal matter or energy? That remains an open question.

THE BULLET CLUSTER

Many scientists consider the Bullet Cluster to be the best evidence for the existence of dark matter. The issue with this cluster is not related to star velocities but to the angle that light is deflected by a group of galaxies. Since I am arguing against dark matter, it is necessary to address the phenomenon of this cluster and see if there are other explanations for the observations astronomers have made that do not invoke the mystery substance.

The Bullet Cluster is the result of one cluster of galaxies shooting right through another cluster of galaxies and creating an interesting pattern of effects. A photo of the collision appears in Figure 9.2. The gas associated with the galaxies has collided, releasing X-ray energy shown in red, but the light-emitting, star-filled galaxies have passed by each other without colliding. The light-emitting galaxies and the inferred associated dark matter are highlighted in blue.

Figure 9.2. The Bullet Cluster

A phenomenon called weak gravitational lensing, whereby light is deflected by gravity as predicted by Einstein, leads to the calculation of the amount of dark matter and its position is shown in blue. The amount that the light deflects is a function of the mass present in accordance with the relationship:

$$\theta = 4G\,M/r\,c^2 \;=\; \frac{4GM}{r^2\,c} \times r/c$$

where (G) is Newton's gravitational constant, (M) is the mass causing the deflection, theta (θ) is the angle of deflection, (r) is the distance from the centre of mass to the incoming light beam trajectory, and (c) is the speed of light. The

DARK MATTER

term (r/c) is the proxy for the time during which the influence of gravity is felt and it has the units of seconds. The mass of the cluster is assumed to be concentrated in a central point from which (r) is measured.

Essentially, the angle of deflection of light indicated by the observations cannot be accounted for using only the mass of the light-emitting matter in the clusters; therefore, dark matter is assumed present with the light-emitting matter.

The dark matter remains with the light-emitting matter and appears to have passed through the collision, emerging along with the galaxies in the clusters, without any interference. Dark matter has the property of being undetectable other than by its gravitational effects. However, it does seem to associate with normal light-emitting matter when mapped out using the statistical techniques of interpreting weak gravitational lensing. I find this to be a curious coincidence in that dark matter could be by itself the dominant component of all matter. Why should it remain with the light emitting component?

A brief description of the gravitational lensing process is in order. Gravitational lenses have no focal length, as shown in Figure 9.3. When light from background galaxies is deflected in passing the mass of the two clusters that form the Bullet Cluster toward the observer, the light converges at the observation point. But the distance of the background galaxies could be anything, and the angle the light gets bent would be indeterminate. Using the red shift of the light from the background galaxies allows physicists to estimate the actual distance to these light sources, and thereby the angle of deflection, and, on with those parameters, calculate the mass that is present.

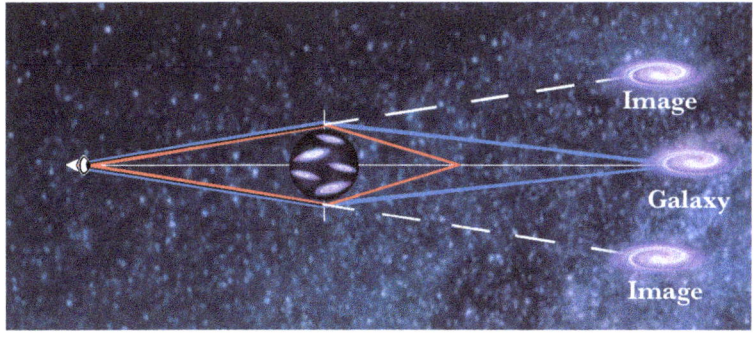

Figure 9.3. Gravitational lensing

There is a relationship between the luminosity of stars and their mass that allows the amount of normal matter in the galaxy clusters to be estimated and compared with the amount of matter required to cause the observed light deflection. The difference, a large mass discrepancy, a factor of five to ten, is accounted for by assuming dark matter is present.

Some scientists think that the difference can be accounted for in other ways, such as the presence of normal matter that does not emit light, like dust, planets, and black holes. The prevalence of black holes in the Universe is not well known, since they are difficult to detect in the absence of inflows of matter that release visible radiation.

Could there be other explanations for the observed deflection of light past the Bullet Cluster? I will offer one possibility here.

The treatment of the cluster as a point mass for purposes of calculating the light deflection angle may introduce a problem. Treating the mass of the cluster as an equivalent point mass is correct mathematically, but it is not necessarily going to predict the real behaviour of the system in practice.

The simplest way to see this is to imagine what would happen if we assumed the mass of the solar system to be concentrated at a point inside the Sun rather than distributed as it really is, among the star and the planets. The true situation allows spacecraft to change their trajectories and get a gravitational energy boost by flying near the other planets. This method is commonly used to alter the course of human-made spacecraft and would not be possible if all the mass was located at a central point.

Similarly, the existence of Lagrange points where there is no net force of gravity because the gravitational tug of one body is exactly offset by the pull of another is not possible if there is only one point mass. Sometimes, convenient simplifications can oversimplify things; one can build a model of a tree out of lumber, but it will not act like a living tree.

Extending this argument to the Bullet Cluster, light passing near individual galaxies near the edge of the cluster could be deflected more substantially than light passing by a point mass at the centre of the cluster at the radius of the entire cluster. The effect could be to give the appearance that there is more mass than is truly present in the cluster. To demonstrate how this might come about, we will do some calculations based on the Milky Way galaxy and the Bullet Cluster. Figure 9.4 illustrates the physical layout.

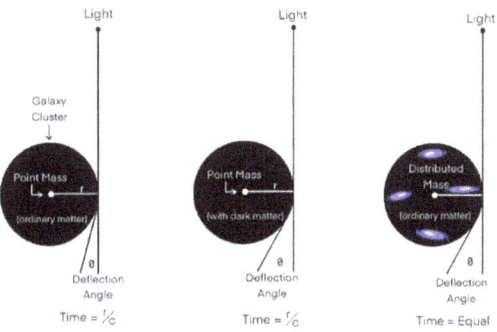

Figure 9.4. Deflection angles vs mass and mass distribution

We begin with the assertion that the observed deflection of light and the distance to the Bullet Cluster and the background galaxies whose light is being deflected are known with some precision. This information is necessary to calculate the mass of the Bullet Cluster using the statistical techniques of weak gravitational lensing. The mass of dark matter seemingly required to achieve the observed bend angle is on the order of six to 16 times the amount of normal matter present.

However, we will attempt to show that the larger bend angle that leads to the assumption of additional mass in the form of dark matter could have a different cause. The force of gravity at the edge of a galaxy can be calculated and compared to the force of gravity on the edge of a galaxy cluster. The cluster has more mass, but also a much greater radius.

We can estimate what these forces are on the equivalent mass of a photon passing near the edge of a cluster of galaxies. A larger force should translate to a larger deflection of the photon.

Table 9.1 contains the results we obtain for the force of gravity, the potential energy, the energy gradient, and the predicted light deflection angle at the edge of the Sun, the Milky Way galaxy, and the Bullet Cluster with and without dark matter.

Using the following relationships, we can calculate the values in the table:

$$F_g = \frac{GMm}{r^2}$$

$$U = \frac{-GMm}{r}$$

$$F_g = k \times \frac{dE}{dr}$$

$$\theta = \frac{4GM}{rc^2}$$

The table contains the basic information about each object. Newton's gravitational constant and the speed of light are the same for all. The mass and radius of each object is different. The calculations apply to a position at the edge of these objects, so the radius is from the centre to the edge. We are assuming that the Milky Way galaxy is a good approximation of the average galaxy in the Bullet Cluster.

The colour-shaded values in the table are used in determining the ratios of the numbers in the corresponding colour value in the ratio column. Notably, the force of gravity at the edge of the Milky Way galaxy is 5.1476 times greater than the force of gravity at the edge of the Bullet Cluster, using the cluster's mass with the assumed dark matter included. The table entries for that calculation appear in blue.

Despite the ratio of gravitational force in favour of the galaxy, the calculated deflection of light is four times greater for the Bullet Cluster than for the Milky Way. The table values highlighted in yellow correspond to this calculation. Intuitively, this does not make sense; more force should mean more deflection.

Parameter	Sun	Galaxy (Milky Way)	Bullet Cluster with Dark Matter	Bullet Cluster without Dark Matter (Mass x 0.25)	Ratio of Parameters in Colored Cells
G (Gravitational Constant)	6.67E-11	6.67E-11	6.67E-11	6.67E-11	
M (Mass in kg)	1.99E+30	6.04E+42	4.97E+44	1.24E+44	4.00E+00
r (Radius in metres)	6.96E+08	3.75E+20	7.71E+21	7.71E+21	
c (Light Speed in metres/sec)	3.00E+08	3.00E+08	3.00E+08	3.00E+08	
Force of Gravity	2.74E+02	2.87E-09	5.58E-10	1.39E-10	5.1476
Potential Energy	1.91E+11	1.08E+12	4.30E+12	1.08E+12	0.2500
dE/dr	2.74E+02	2.87E-09	5.58E-10	1.39E-10	5.1476
Theta (radians)	8.49E-06	4.79E-05	1.91E-04	4.79E-05	0.2500
Theta (arc seconds)	1.75	9.87	39.50	9.87	0.2500
Equal Time Theta (radians)		9.86E-04	1.91E-04	4.79E-05	5.1476
Equal Time Theta (arc seconds)		203.32	39.50	9.87	5.1476

Table 9.1. Light deflection angle and force of gravity

The reason for the discrepancy is in the way the equation for calculating the deflection angle theta (θ) is derived. The time that the force is presumed to act is given by the term (r/c), the radius of the object divided by the speed of light, in units of seconds. This term is imbedded in the equation for theta. If we instead allow the time to be equal in each case, we find the results are consistent with the ratio of gravitational forces and the bend angle is 5.1476 times greater for the Milky Way galaxy than for the Bullet Cluster with dark matter included.

We then adjust the presumed mass of the Bullet Cluster with dark matter downward, essentially removing the dark matter to get the bend angle for the cluster to agree with the standard theta calculation for just the Milky Way. To do this, we multiply the mass by a factor of 0.25 times to get the required matching angle of 4.79E-5 radians.

In other words, the Bullet Cluster needs a mass four times greater than what seems to be present in light-emitting matter to bend light as much as a single galaxy like the Milky Way would using the conventional method of calculating the bend angle theta. If we use the equal time approach, the Milky Way-like galaxy deflects the light even more, in fact 5.1476 times more in terms of the angle theta.

The conclusion from these calculations is that light passing near a galaxy located near the fringe of the Bullet Cluster should be deflected at an angle

5.1476 times greater than the angle if it had passed by the point mass of the cluster with dark matter included at the radius of the cluster.

The gravitational effect of the remainder of the Bullet Cluster can be superimposed on that of the galaxy on the edge of the cluster represented by the Milky Way to give a deflection of approximately six times that of the whole cluster. This figure corresponds reasonably well with the assumption that 85% of the mass of the Bullet Cluster must be dark matter: 85%/15% = 5.67.

The issue with using equal times instead of (r/c) to calculate the deflection angle theta would not have been noticed in the case of the Sun, a singular mass. The version of the equation for theta that incorporates (r/c) gives the correct observed deflection of 1.75 arcseconds for light passing near the Sun. That deflection confirmed Einstein's theory of relativity in 1919, as the eclipsed Sun passed in front of the stars of the Hyades cluster and caused their light to be deflected by that angle.

Assuming a cluster is a point mass, however, is a different situation, with problems analogous to assuming the solar system is a point mass discussed previously.

Using equal times in the calculation of the bend angle theta, instead of using (r/c), brings the bend angle into agreement with the force of gravity calculation and is, therefore, thought to be more reasonable.

The bottom line is that the Bullet Cluster does not necessarily confirm that dark matter exists. The approach discussed here renders the extra mass unnecessary to explain the observations.

This concludes the argument against dark matter. Should dark matter ever be found and identified, then this argument must be flawed, and the correspondence of the results found here to the accepted ratio of dark matter to normal matter must be just coincidental.

AS MERCURY TURNS

Another example of deflection due to relativistic effects is the precession of Mercury's elliptical orbit around the Sun. The orbit of Mercury was found to have apsidal precession, an effect where the direction of the major axis of its elliptical orbit changes by a very small amount with each revolution the planet makes around the Sun. The amount of precession observed was always about 43 arc seconds more per century than the 531 arc seconds that could be accounted for by the classical gravitational effects of the other planets.

One of the three major tests of Einstein's relativity (light deflection by mass, and gravitational redshift being the other two) was a relativistic adjustment to the calculated precession of Mercury's orbit. Einstein was able to predict the observed behavior of the planet perfectly, within observational error.

When a body such as a planet orbits a star in a perfect circle, its orbital velocity remains constant. However, if the orbit is elliptical there is a reciprocation of gravitational potential energy and kinetic energy. The velocity of the planet changes in accordance with Kepler's laws of planetary motion. In summary these three laws are: 1) The orbit is an ellipse with the Sun at one focus, 2) a line joining the planet and the Sun will sweep out equal areas in equal time intervals, and 3) the square of the orbital period is proportional to the cube of the length of the semi-major axis of the orbit.

Planets orbiting the Sun vary in their velocity, speeding up as they come closer to the Sun, and slowing as they move further away. This change in velocity would also cause a rotation into or out of the energy dimension, as outlined earlier. A speculation is therefore offered at this point, that the precession of the orbit of Mercury is related to this phenomenon. The orbital trajectory in the space dimensions is reduced in correspondence to an increase in energy dimension quantified by the (v/c) parameter in the Lorentz transform. Equivalently, we could consider that Mercury's relativistic mass increases with its velocity, resulting in a stronger classical gravitational force that makes it corner more sharply around the Sun than it would without relativistic effects.

This description of how the space-energy hypothesis addresses the precession of Mercury's orbit is presented for consistency with this important test of relativity. The phenomena of gravitational red shift and light deflection will be explored further in later chapters in the context of variable light speed and evidence for dark matter respectively.

DARK ENERGY **10**

DARK ENERGY

Current evidence suggests that the expansion of the Universe is accelerating. This is a quandary for science since gravitational forces should be slowing the expansion rate down. Like the cosmic inflation postulate that has the cosmos suddenly expanded at incredible rates a smidgen of a second after the Big Bang, dark energy is assumed to cause this later-life increase in the rate of expansion. Dark energy, like whatever supposedly caused cosmic inflation, is an unknown form of energy. A complete unknown yet estimated to account for 68% to 71% of the of the total energy in the Universe today.

Dark energy is one explanation for the observations, but is it the only possibility?

How do we know the Universe is expanding more quickly lately? The news comes on the redshifted light from supernovae, exploding stars. Because supernovae of Type 1 are consistent standard candles of known intrinsic brightness (absolute magnitude), we can estimate their true distance from Earth by observing their brightness (apparent magnitude) and doing some calculations.

The light from these explosions can be analysed to find their redshift. Redshift is designated by the letter Z and corresponds to the change in wavelength due to relative motion of the source with respect to the observer as well as the change due to the expansion of the Universe itself. Therefore, we can estimate the recession velocity of objects from redshift and their distance from their apparent brightness, and then determine if there is a trend.

Edwin Hubble (1889–1953), an American astronomer, initially established that the Universe is expanding by employing the redshift of Cepheid variable stars, whose period of cyclical brightness variation is related to their absolute magnitude. Supernovae are much brighter than Cepheid variables, so they can be used for the same purposes over much larger distances.

In 1998 the Supernova Cosmology Project and the High-Z Supernova Search Team concluded that the expansion of the Universe, rather than slowing, is accelerating. The 2011 Nobel Prize in Physics was awarded to members of these research teams: Saul Perlmutter, Adam Riess, and Brian P. Schmidt.

Accelerating expansion can be explained mathematically by Einstein's cosmological constant, a term in his equations that he called his "greatest blunder." This term in the equation offsets the gravitational collapse, drives expansion, and is now considered to represent the cause of accelerating expansion: dark energy.

$$ R_{\mu w} - \frac{1}{2} R g_{\mu \nu} + \Lambda g_{\mu \nu} = \frac{8 \pi G}{c4} T_{\mu \nu} $$

Einstein's equation contains tensors that describe how mass and energy are distributed and how space-time curves in response. Tensors are mathematical expressions represented by an array of components that vary with position in space. The equation contains the cosmological constant term (Λ). The term ($R_{\mu\nu}$) is the Ricci tensor, (R) is the Ricci scalar, ($g_{\mu\nu}$) is the metric tensor, (G) is Newton's gravitational constant, ($T_{\mu\nu}$) is the stress-energy tensor, and (8π) is what I did for dessert after dinner the other day.

One significant possible test for the space-energy hypothesis is whether a mathematically equivalent form of these equations can be generated that replaces time with an energy-based substitute. We have seen how Newton's ($F = ma$) can be expressed without time.

Understanding and explaining dark energy is one of the major challenges in front of modern cosmology. So, here is an attempt to do that.

In an earlier chapter, we saw how the variation of the speed of light in accordance with the equation,

$$c = \sqrt{\frac{E}{m}}$$

could occur. With an ever-declining mass and energy density, inverse of an ever-expanding volume, the speed of light and expansion should be slowing.

A history of the Universe appears in Figure 10.1. The Universe evolves from left to right in this graphic. Note the rapid early expansion, a result of extremely high energy density at the outset, thought to be due to an unexplained cosmic inflation energy. Also note the steeper broadening of the expansion "bell" in more recent times, thought to be due to dark energy.

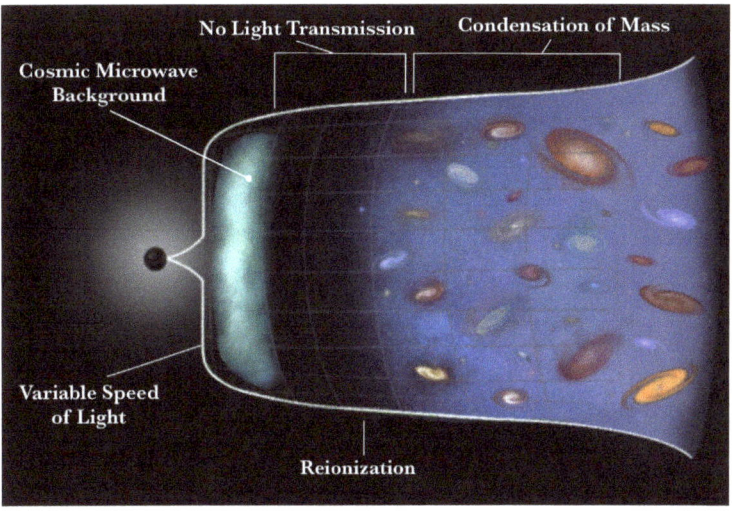

Figure 10.1. Accelerating Expansion

The hypothesis is now submitted that both effects—early rapid expansion and late accelerating expansion—can be explained by the variable speed of light paradigm. The chapter on light speed discussed cosmic inflation at the beginning of the life of the Universe, so here we will focus on the recent accelerating expansion.

Again, the variable speed of light hypothesis, as developed from energy and mass densities of the Universe, would suggest that the expansion should slow as the speed of light slows. Expansion of the Universe is taken to be the progress of light outward from the Big Bang, at the prevailing speed of light over the course of its evolution.

REDSHIFT

The evidence from observed redshift data seems to indicate accelerating expansion, but could it just be an artifact of the observations? This is not to suggest any errors in measurement or uncertainty with the techniques used. The observations are likely to be correct, within present experimental precision. The redshifts are real.

The redshifts are real but may have two superimposed components. There is the normal Doppler effect redshift that results from the relative velocity of source and observer. The ambulance siren analogy where the sound pitch of an approaching source increases and the pitch of a receding source decreases explains the Doppler component. Blueshift and redshift are the equivalent of the higher and lower pitch shift in the case of light.

Another possible component results from the variation in the speed of light. Imagine an ambulance approaching at extremely high velocity, slowing down, passing the observer, then moving away but accelerating to very high velocity again. Would the observed siren pitch not change in a different manner than if the velocity of the ambulance were assumed to be always constant?

Calculations show that there are two components of redshift if we assume an accelerating sound source. We can mathematically quantify the Doppler shift using the equation,

$$f_o = f_s \left(\frac{v + v_o}{v + v_s} \right)$$

where (f_o) is the frequency the observer hears, (f_s) is the frequency of the source, (v) is the speed of sound, (v_o) is the speed of the observer, and (v_s) is the speed of the source.

Doppler Effect

Frequency Observed (Hz)	Frequency of Source (Hz)	Speed of Sound in Air at 20 C (m/s)	Velocity of Observer (m/s)	Velocity of Source (m/s)	Frequency Observed (Hz)	Frequency of Source (Hz)	Speed of Sound in Air at 0 C (m/s)	Velocity of Observer (m/s)	Velocity of Source (m/s)
440	440	344	0	0	440	440	332	0	0
416	440	344	0	20	415	440	332	0	20
394	440	344	0	40	393	440	332	0	40
375	440	344	0	60	373	440	332	0	60
357	440	344	0	80	355	440	332	0	80
341	440	344	0	100	338	440	332	0	100

Table 10.1. Doppler effect

The table shows how frequency observed varies when the velocity of the source changes with respect to the observer in the case of a 440 hertz ambulance siren in 20 degrees Celsius air. Note that if an ambulance moves away from the observer at a constant velocity of 20 metres per second, the true 440 Hz siren frequency is heard at 416 Hz. If it accelerates to 40 m/s, the frequency decreases to 394 Hz. At 100 m/s, 341 Hz is heard by the observer. These changes in frequency are the result of a change in the velocity of the source.

What if we introduce a change in the speed of sound through the air? We would probably not be inclined to believe that the speed of sound in air was changing, although it does with air temperature and density. Let us consider the case where the source is moving away at 100 m/s and look at the frequency observed if the speed of sound changes from 332 m/s to 344 m/s, as it does when air warms up to 20 degrees Celsius from 0 degrees Celsius. Now a true frequency of 440 Hz at the source is observed at 341 Hz instead of 338 Hz. The pitch is not lowered quite as much.

Were we considering light in the above example instead of sound, we would say the light is less redshifted instead of the pitch being lowered by a smaller amount. As we look further back in time, further out in space, we see that the redshift becomes less than we would expect if we assumed the speed of light has always been constant. This corresponds to the appearance that the expansion of the Universe is accelerating. Figure 10.2 shows graphically the departure of the speed of expansion of the Universe from the linear Hubble relationship as we look further away and further back in its history.

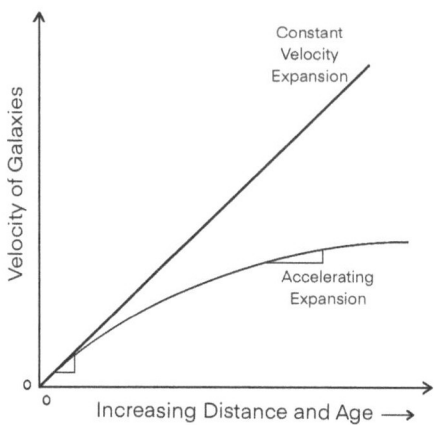

Figure 10.2. Accelerating expansion of the Universe

If we assume that the speed of sound is always 332 m/s, we will interpret the pitch change from the expected 338 Hz to 341 Hz as a slowing of the ambulance, a deceleration as it moved further away, from 100 m/s to about 96.5 m/s. In the case of light, we would see less than expected redshift as we looked back further in history and conclude that the expansion of the Universe must be accelerating in more recent times. We would have to invoke some sort of dark energy to cause this acceleration, but the actual cause might be a change in the speed of light.

Time for an analogy to investigate the imprinting of a frequency onto a carrier. Let us spin a bicycle wheel with a one-metre radius at a certain angular velocity, say 60 revolutions per minute, one revolution per second, therefore a frequency of 1 Hz. First, though, we put a mark on the wheel that we can see and trace its position. Now let us move the wheel, suspended on an axle as a carrier, not rolling on the ground, and track the mark on it. The trajectory the mark follows is a sine wave.

If we move the wheel horizontally on a line perpendicular to the axle, at 6.28 or (2πr) metres per second, the mark will create a sine wave with a wavelength of 6.28 metres. If we double the horizontal velocity the wavelength will be 12.56 metres. We have created a longer wavelength, a redshift, by changing the velocity of the carrier.

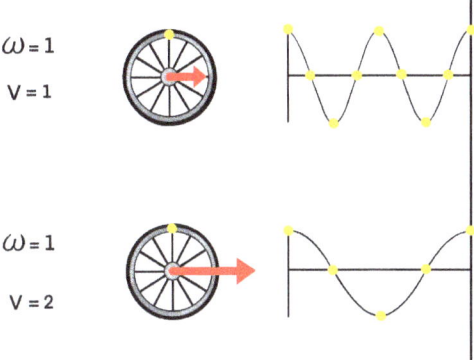

Figure 10.3. Wave created by the rotation of a wheel

In the bicycle wheel analogy, what remains constant is the angular rotation rate of the wheel, the energy content, that is directly proportional to the frequency of light waves at constant carrier speed, by the equation,

$$E = hf$$

where (E) is the energy content, (h) is Planck's constant, and (f) is the frequency of the light.

We also know that the frequency is dependent on the speed of light,

$$f = \frac{c}{\lambda}$$

For the energy content of a photon to be conserved, the frequency must remain the same, and therefore, if the speed of light is faster, the wavelength must increase in proportion, as we see with the bicycle wheel. Light behaves this way when it enters a refractive medium such as water; both the speed and the wavelength change in proportion.

Like the forward velocity of the bicycle wheel axle, the speed of light is presumed to be changing in accordance with the space-energy paradigm. Light is the carrier of energy, which is related to the spin rate or frequency of the sine wave. Most of the change in the speed of light from the Big Bang to the present occurred as a dramatic reduction very early on, during the period of cosmic inflation. More recently the change has been relatively modest, as can be seen in Figure 10.4, which originally appeared in Chapter 6.

Also shown in Figure 10.4 is a graph of the observed Z factors throughout the life of the Universe for comparison. The observed Z factor relationship bear an uncanny resemblance to the calculated speed of light based on the declining energy density as the cosmos expands. This perhaps lends some credence to the variable speed of light idea.

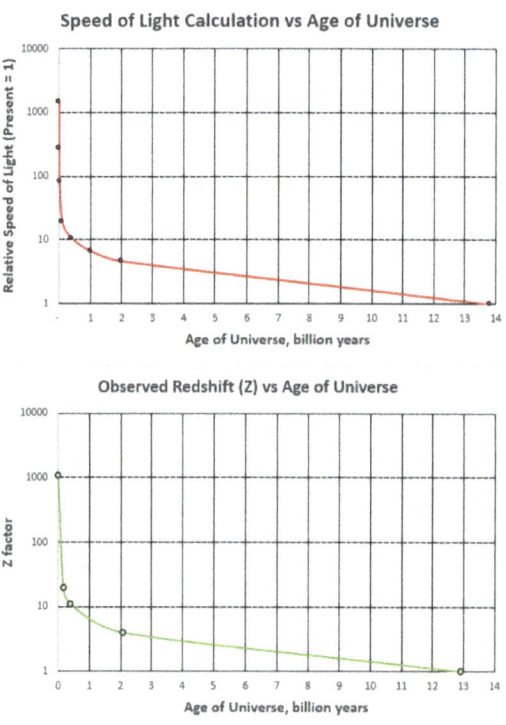

Figure 10.4. Speed of light and Z factor vs time

DARK ENERGY

In terms of the history of the Universe, the expansion was much faster earlier on, further back in time, back as far as we can see, to the cosmic microwave background. We can only see stars back to the time when energy began to condense into mass and ions formed atoms, and the first stars formed. The space-energy hypothesis proposes that the speed of light was faster earlier in the life of the Universe, and that mass that condensed out earlier would have been moving faster as well. Matter in space only moves at a small fraction of the speed of light, 1% or less. Thus, the stars and galaxies that formed earlier may be moving away faster through space and consequently show more standard Doppler redshift.

Superimposed on the frequencies imprinted on the carrier is the cosmological redshift that results from the change in the speed of light. When we look back as far as we can, to the Cosmic Microwave Background (CMB), the remnant radiation from the Big Bang, we see it is redshifted with a Z factor of about 1089. The temperature of the CMB is about 2.73 degrees Kelvin or -270 degrees Celsius, reduced significantly from the 3000 degrees Kelvin temperature it had roughly 380,000 years after the Big Bang, when hydrogen atoms could form, and space became transparent to light.

If the speed of light were indeed much higher then, the wavelengths imprinted on it would be much longer, and that is what is observed. This cosmological redshift, that puts the wavelength of the CMB radiation in the range of 1 millimetre, as opposed to 1 micrometre at 3000 degrees Kelvin, is attributed to the stretching of space itself as the Universe expands. However, the same effect can be obtained if we utilize the bicycle wheel analogy: the wavelength is stretched if we move the carrier faster and assume that the energy is directly associated with the angular velocity of the wheel itself.

The implication of the variable speed of light idea is that the speed would have been approximately 1089 times faster than it is today at the time when atoms first formed, 380,000 years after the Big Bang. Referring again to Figure 10.4, we see a close to 1000 ratio between the initial speed of light and today's value, but the timing is off. Unless we count the expansion of the Universe from the time that light was first able to propagate freely, which seems a reasonable adjustment.

The Universe itself is postulated to expand at the contemporaneous speed of light. The faster speed of light in the beginning accounts for the radius of the Universe being 46 billion light years, although the observable portion has a radius of only 13.8 billion light years.

LIGHTING UP THE DARKNESS

To wrap up, dark energy within the context of the space-energy model is not necessary. The expansion of the Universe is proposed to simply be the rush of ordinary energy away from the Big Bang at the prevailing speed of light, and it has been slowing down. This expansion might appear to be accelerating, based on observations of higher redshift in the light from objects that were formed earlier on when the speed of light was much higher, creating the false impression that the expansion is accelerating.

In the model proposed in this book, the Universe continues to expand at the prevailing speed of light, and that speed is slowing down. No acceleration, no dark energy.

BLACK HOLES

11

COSMOLOGY—BLACK HOLES

Black holes are the object of much scientific interest these days. Their existence was predicted mathematically by Einstein in his general theory of relativity in 1915, though he was reluctant to believe they existed physically. In 1916, Karl Schwarzschild (1873-1916) worked out a solution to Einstein's equations that predicted any mass, if compressed into a sphere of less than a certain radius, would have a gravitational attraction so strong that even light could not escape. Essentially the escape velocity required to leave a black hole is equal to the speed of light, and nothing can exceed that speed.

This radius that a mass must be compressed into, for any mass (M) is now known as the Schwarzschild radius (r_s), and is defined as shown below:

$$r_s = \frac{2GM}{c^2}$$

Newton's gravitational constant (G) appears again. Very notable is the dependence of the Schwarzschild radius (r_s) on the speed of light (c), in the context of a VSL theory. If we accept the premise that there is nothing special about the current value of the speed of light and it can vary, then the equation tells us the Schwarzschild radius can vary also. The Schwarzschild radius is the event horizon radius, the distance where escape from a black hole becomes impossible.

The much higher speed of light, postulated in the early life of the Universe, would make the Schwarzschild radius smaller, requiring more concentration of mass to create a black hole. This might suggest that black holes became more common at some point as the speed of light slows, expanding the Schwarzschild radius sufficiently to ease black hole formation. If mass is distributed in space at a density less than required to form a black hole, but the Schwarzschild radius expands to encompass more mass—would that trigger the formation of a black hole?

The preponderance of black holes during various phases of the evolution of the Universe could be investigated to determine whether there is any relationship. We are hampered by the fact that black holes are difficult to observe in the absence of mass falling in and releasing radiation around the hole, beyond the event horizon, since they cannot emit light themselves.

The first stars are thought to have appeared around 100 million years after the Big Bang. Galaxies became common around one billion years, although the most recent data from the James Webb Space Telescope (JWST) indicates the presence of large galaxies even earlier than expected. The hypothesis of this book may be consistent with earlier formation of stars and galaxies because it allows for steeper gradients in the energy background early on when the energy density was greater, hence stronger gravitational forces. Stronger gravitational forces should accelerate formation of mass accumulations allowing formation of structures in less time than anticipated by conventional theory. At the time of this writing, the new JWST observations are causing much consternation among scientists. The entire Big Bang theory is being called into question.

Quasars, or quasi-stellar objects, emitting immense radiant energy are thought to be powered by gas falling into black holes at the centre of galaxies that formed during the first few billion years of the life of the Universe. Quasars may be among the most distant objects from the Earth, with the most distant of them being about 13 billion light years away, near the edge of the observable Universe at 13.8 billion light years.

Black holes appear like cosmic drains at the centre of galaxies. Galaxies appear like swirls of orbiting stars circling the drain. In a sense, black holes do drain mass and energy from the Universe that we know, disappearing it inside their event horizons. Once something crosses the event horizon, it effectively loses contact with the external Universe because no light or information can escape. Black holes are characterized by their mass, their charge, and their spin. What happens inside a black hole is a matter of intense speculation, and like Las Vegas, what happens in a black hole does stay in the black hole.

Black holes absorb both mass and energy that they encounter, but can be characterized by their mass alone, although charge and spin are required for complete description. Essentially, black holes convert mass and energy into a mass equivalent, since any energy inside cannot escape.

Any mass has a mathematical Schwarzschild radius, from subatomic particles to the Universe itself. Interestingly, the Schwarzschild radius of the Universe calculates as 23.5 billion light years, using the estimate for ordinary mass-energy content of 1.5×10^{53} kg. This suggests that if all the mass were compressed into that radius, we would be living inside a black hole.

The complete radius of the Universe is said to be around 46 billion light years, but we can only observe back in time to when atoms formed, and radiation became able to propagate freely. We can observe the cosmic microwave background (CMB) that is the leftover radiation of the Big Bang. We can see photons more recently emitted by objects out to the distance that light is now arriving from at its current speed.

Note though that under the variable speed of light hypothesis light has always travelled at the prevailing speed of light, initially much faster than the present-day value. This means that light from the very edge of the Universe, 46.5 billion light years could have reached us if the Universe had always been transparent.

PIGEONS ACROSS THE UNIVERSE

A pigeon analogy seems perfect to clarify the situation. Assume you have a crate of homing pigeons that can carry messages back to their loft, but they are limited to flying at 60 kilometres per hour. Take the crate of pigeons in a truck and drive them away at 120 km/hour. When you are 60 km from home, in half an hour of driving 120 km/hour, you release one with a message stating your release point. In another hour that pigeon should arrive back at the loft, that is, in an hour and a half total elapsed time since you left home. Keep driving until you are 120 km away, in one hour, then release another pigeon. That one will arrive in three hours total elapsed time.

Now, assume an observer is waiting at home for the messages to arrive. After one and a half hours, the first pigeon arrives, representing his observable horizon, 60 km away at that time. There is another pigeon coming from 120 km away, but it cannot be seen at present, as it is outside his observable horizon. This assumes constant pigeon airspeed velocity. If, however, the pigeon could initially fly much faster, and then later slowed to 60 km/hr, due to tiring out or picking up a coconut to carry,[3] in principle a pigeon from much further away could be seen after an hour and a half.

The discrepancy of 13.8 to 46 billion years is conventionally explained by having space itself expand faster than the speed of light. The theory of this author is different, in that the speed of light was initially much higher than it is today, and the universal expansion has occurred at the prevailing speed of light throughout its entire lifespan. So, the result is light has been able to travel 46 billion light years in 13.8 billion years. A recent study published in 2023 has put the age of the Universe at 26.7 billion years by studying the oldest stars and finding them to be older than 13.8 billion years, but this, too, might be an artifact of a variable speed of light.

In the pigeon analogy, the speed of the pigeon is the speed of light today. The speed of the truck is the speed of light in the past. The actual variance is a smooth curve, not a two-value step function, which was just a stool pigeon for illustration purposes.

If the hypothesis presented here is correct, the speed of light will continue to slow down, and the Schwarzschild radius of the Universe will continue to expand. More will be said about this in the chapter on the fate of the Universe.

---

3   This is a reference to the Monty Python movie *The Holy Grail*, where swallows can be laden with coconuts, which affect their airspeed velocity. If you have not seen the movie, this will make no sense at all, proving conclusively that everyone should see the movie!

NO SINGULARITIES

We do not know what occurs inside a black hole, but the argument will now be made that it is not a singularity. A singularity is a dimensionless point into which all the mass would be compressed. Singularities are a mathematical possibility but probably not a real physical thing. Most physicists no longer believe in singularities, but it is worthwhile to present the author's logical justification for that disbelief.

As discussed in an earlier chapter, inside a solid spherical mass, gravity behaves differently than outside the boundary of the mass. The force of gravity is still described with Newton's relationship,

$$F_g = \frac{GMm}{r^2}$$

but for an observer located inside the sphere, there is a smaller sphere of mass closer to the centre and a hollow spherical shell of mass further away. The force of gravity varies linearly with the distance from the centre and is proportional to the mass within the radius that the observer is stationed, as envisioned in Figure 11.1.

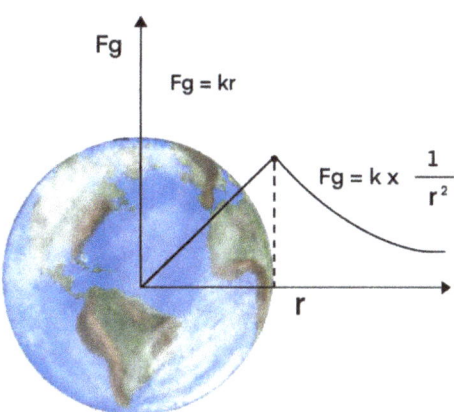

Figure 11.1. Force of gravity inside and outside a body

The mass (M) within a radius less than the surface radius is, for a spherical object like a planet, star, or black hole, given by

$$M = \frac{4}{3}\pi r^3$$

where (ρ) is the average mass density of the object.

Combining and cancelling like terms, we get

$$F_g = G\left(\frac{4}{3}\rho\pi r^3\right)\frac{m}{r^2} = G\rho\left(\frac{4}{3}\pi m r\right) = kr$$

where (k) is a constant combining all constants.

Consequently, the force of gravity is zero when the radius from the centre is zero. At the centre of the Earth, or any spherical mass, there is no gravitational force, as all the surrounding mass pulls away in every direction.

We combine this finding about internal gravitational force with the notion that energy cannot escape from a black hole. The mass inside a black hole has potential energy in relation to its distance from the centre and, to become compressed into a point, would have to lose that energy somehow, and that cannot happen because nothing can escape. And thus, no singularity!

The average density of a black hole declines as it grows because the radius increases proportionally to mass while the volume increases with the cube of the radius. But perhaps at the center of a black hole is a very dense core of mass that under certain circumstances converts directly to energy, but that is sheer speculation.

Black holes could play a very important role in the Universe in a variety of ways. Their mass at the centre of galaxies could affect the galactic mass distribution and, accordingly, could explain the galactic rotation curves we see. If black holes are distributed throughout galaxies, in a spectrum of different sizes, the mass distribution might also be significantly affected. Recent observations of gravitational waves indicate there is a very low frequency thrumming coming from massive black holes throughout the Universe.

Although it is just speculation, within the context of the space-energy model, where mass causes a depletion in the energy background, black holes might represent the extreme case. The energy background intensity might be reduced to zero within the event horizon. That is not to say there is no energy inside black holes, just that there may be no background energy. Two analogies will be called upon to explain this idea, and since we have used everything but the kitchen sink . . .

Take the water in a kitchen sink. Around the drain, the water level is drawn down, even beyond the water depth of the sink bottom, into the plumbing below.

In the atmosphere, a tornadic funnel cloud can develop, but they do not always touch down to the ground. When they do meet the ground, they cannot go any lower, but they can grow in diameter. Hurricanes also have an eye at the center where the pressure is at a minimum. To emphasize a point made above about energy inside a black hole, there is air inside a funnel cloud and the eye of a hurricane.

Imagine if the Schwarzschild radius represented a withdrawal from the energy background of all that was available. Not just the dimple in the energy background that normal objects like stars and planets and coffee cups cause, but a complete excavation that can go no deeper, only grow in radius. Black holes grow their Schwarzschild radii as more mass falls in.

In some ways, black holes are the opposite of stars, taking mass and energy out of space and sequestering it. Stars do the reverse: they convert mass into radiant energy and shine it out into the cosmos. The fact that black holes can be described with just three parameters—mass, charge, and spin—makes them consumers of information, draining the complex information required to fully describe galaxies, stars, planets, lumps of rock, gases, dust, and life away inside their event horizons, rendering it inaccessible. They can also be regarded as sequestering energy such that it only appears as mass from the outside.

GRAVITATIONAL WAVES 12

"It is obvious that the amplitude of gravitational waves has,
in all imaginable cases, a practically vanishing value."

— **Albert Einstein (1879–1955)**

Gravitational waves are fluctuations in the intensity of gravity that travel at the speed of light. They were first proposed by Oliver Heaviside (1850–1925), an English mathematician and physicist who made many contributions to science. In 1905, Henri Poincaré (1854–1912) likened gravitational waves to electromagnetic waves. Then in 1916, Albert Einstein predicted gravitational waves as ripples in the space-time fabric of the cosmos, but later rejected them as real physical phenomena and remained skeptical about them to the end of his life.

Newton's law of universal gravitation assumes the effects of gravity are instantaneous, travelling at infinite speed. Even Newton was at a loss to explain how the force of gravity could transmit without contact between objects in the absence of an intervening media. Of the mechanism for this mysterious "action at a distance," Newton famously commented *"hypotheses non fingo,"* Latin that translates roughly as, "I have no idea."

This conundrum gave rise to the idea of some sort of field that fills all of space. James Clerk Maxwell (1831–1879) treated electricity and magnetism as fields and developed the electromagnetic field theory that linked the two and illuminated the fact that light is an electromagnetic wave that propagates at its eponymous speed.

We now know by experiment that gravitational waves do propagate at the speed of light, and they can be and have been detected.

The first indication of gravitational waves was obtained by observing a neutron star and a pulsar in orbit around their common centre of mass. This dual star system is the Hulse-Taylor binary, named in 1974 for its discoverers, Russell Alan Hulse and Joseph Hooton Taylor Jr. of the University of Massachusetts Amherst. The orbital decay of these two stars was exactly as predicted by the general theory of relativity, in agreement with the expected loss of energy due to gravitational radiation.

In 2015 gravitational waves were detected directly when the merger of two black holes sent a gravity signal detected by the Laser Interferometer Gravitation Wave Observatory (LIGO) detectors located in the United States. This discovery resulted in the 2017 Nobel Prize in Physics going to Rainer Weiss, Kip Thorne, and Barry Barish. The LIGO consists of test masses located along two orthogonal arms that can measure, using laser interferometry, any motions between these free masses caused by gravitational waves. Relativity theory ascribes this movement to the stretching or distortion of the fabric of space-time. The alternative proposed here is that the motion is caused by the passage of an energy density wave through the energy background. The energy gradients associated with the density wave produces a force on the masses as it passes through, as illustrated in Figure 12.1.

Just as mass creates an energy deficit and consequently, energy gradients responsible for the force of gravity, a passing gravitational wave creates very small energy gradients associated with the wave. The force that results from a wave is related to the waveform. You will recall from the chapter on force that all forces are due to energy differentials or gradients.

Forces from Energy Waves

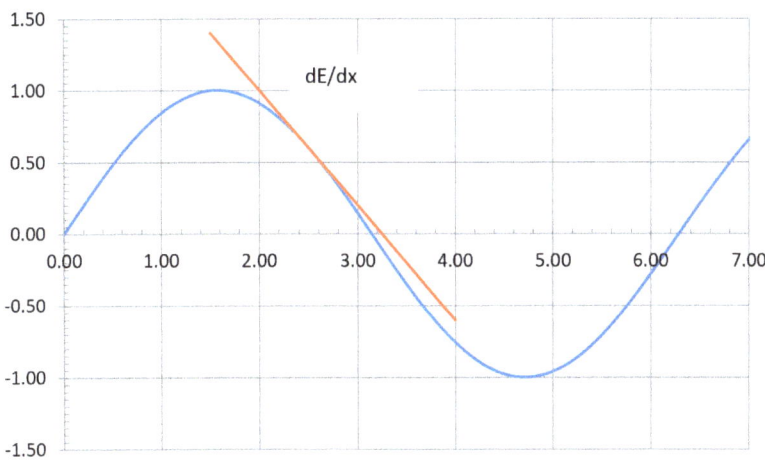

Figure 12.1. Forces from energy waves

The theory of gravity expounded in this book, as you know by this point, is an energy background throughout the Universe that is distorted or dimpled by the presence of mass. The energy background is composed of electromagnetic radiation that was present at the Big Bang and persists since mass-energy can neither be created nor destroyed. More precisely, the background is the total energy of the Big Bang, less that amount that has condensed out as mass, plus the amount generated by stars, less the amount that has been absorbed by black holes, and relatively minor additions or subtractions caused by all other phenomena. Disturbances within this media travel at the speed of light.

The Universe tends to maintain a mass-energy distribution, just as the air in a balloon statistically maintains a constant pressure throughout its interior, and like the water in a lake, maintains a constant surface level (neglecting the curvature of the Earth) in the absence of a wave driving force such as the wind. With the Universe, the distribution is a little more complicated than a level surface, since mass creates spherical deficits in four-dimensional space-energy, so it might be more accurate to say an overall equilibrium is maintained that is consistent with this pattern.

ENTROPY

Gravitational waves travel out from a gravitational disturbance and dissipate, as do the waves in water when created by, for instance, the stirring of a boat motor propeller. Energy seeks a level, a state where there are no energy gradients, thus no forces. This Universal tendency is the property we call entropy, or at least it is, by definition, in the space-energy paradigm. Entropy rules!

The classical definition of entropy is a thermodynamic quantity representing the availability, or lack of availability, of the thermal energy of a system for conversion into mechanical work. Mechanical work is force times distance, and with no energy gradient, there is no force, therefore no work is possible. A uniform energy background with no density variance corresponds to a maximum entropy situation. The intensity of the energy does not matter, only the uniform distribution with no gradients. Normally, it is assumed that this state occurs at the beginning and the end of the Universe.

Entropy is often interpreted as the degree of randomness or disorder in a system. Is a uniform energy background more random and disordered than one with density waves travelling through? Certainly, there is more information in a non-uniform energy background if it takes more information to accurately describe it. The dissipation of gravity waves would directionally lead back to a more uniform background.

WAVES

Gravity waves are a natural consequence of the space-energy model. A medium for waves to propagate through is present, and the speed of light is precisely how rapidly energy transits through that media.

The gravitational forces are, of course, very weak, their strength being inversely proportional to the square of the distance from their source, just like gravitational potential energy. When two large masses come together, such as two black holes spiralling rapidly toward one another, the two associated gravitational wells move rapidly and then merge into one. This creates an intense disturbance that propagates out through the energy background in the form of waves. The situation is not unlike the ripples on a lake; a stone, or in the case of merging black holes, two stones, thrown into the water make waves.

Gravitational waves transport energy as gravitational radiation, a form of radiant energy that is thought to be like electromagnetic radiation. Gravitational waves are transverse waves, just as electromagnetic waves are—that is, they oscillate in directions transverse to the direction they propagate in.

Let us look at an analogy using air in a room. The air molecules move in random directions at a certain rate that relates to the temperature of the air. Sound can travel through this very same air, but sound is a pressure disturbance superimposed on the molecular motion, as shown in Figure 12.2.

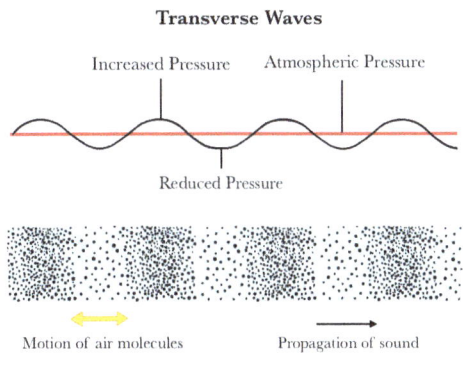

Figure 12.2. Waves propagating through air

In the energy background, gravitational waves are disturbances in the total electromagnetic field density, and the field itself is composed of individual photons or electromagnetic waves. The two different effects are superimposed. However, unlike the longitudinal compression waves in air, gravitational waves are energy density waves.

In conclusion, we wave goodbye with the unwavering belief that gravity waves fit well with the model that provides a media for them to exist in.

MASS-ENERGY

13

"Energy cannot be created or destroyed; it can only be changed from one form to another."

— **Albert Einstein (1879–1955)**

"Reality is made up of circles, but we see straight lines."

— **Peter M. Senge (1947–)**

MASS-ENERGY

Mass and energy are interchangeable. They are possibly different phases of the same thing, like water and ice are different phases of the H_2O molecule. Mass can be converted to energy; the quantity of energy theoretically available equals the mass multiplied by the speed of light squared. While mass can never achieve the speed of light, because of the infinite amount of energy it would take to accelerate to that velocity, massless electromagnetic energy automatically travels at the speed of light. This begs the question of how mass, in converting to energy, can go from a standstill to the ultimate speed limit instantaneously? Does this not imply an infinite acceleration?

One possible solution to this paradox is that energy locked up in the form of mass continues to move at the speed of light, but it does so in a cyclic way, analogous to moving along a closed path like a circle. Another concept is that energy is a propagating electromagnetic wave and mass is a corresponding standing wave. In quantum mechanics, the wave function of confined particles such as electrons have wave functions corresponding to standing waves.

Electromagnetic waves encountering a mass may react in various ways: they may be reflected, transmitted, or absorbed. Absorption is the only outcome where the wave becomes part of the mass and will only occur at certain frequencies that the electrons in the matter will accept by making a quantized jump to a higher energy level. However, complex atoms and molecules, beyond simple hydrogen with one electron orbiting one proton, can absorb a wider range of frequencies and increase their temperature (vibrational speed) or internal energy level.

If the difference between mass and energy is, in fact, the difference between propagating electromagnetic waves and standing electromagnetic waves, then the relationship is a little clearer. In four-dimensional space-energy, the distinction is between waves propagating in the energy dimension versus waves pinned to a locality in the three dimensions of space.

A simple two-dimensional analogy would be a traffic circle or roundabout, as shown in Figure 13.1. A straight highway meets up with a circular roadway with many entrances and exits, where vehicles can change from linear motion to circular while maintaining their speed, assuming they are travelling reasonably slowly. Electromagnetic energy travels along straight lines, like the roads, until it becomes confined in an orbit or standing wave pattern represented by the traffic circle.

Figure 13.1. Traffic circle analogy

The centripetal force required to make a mass follow a circular path is given by,

$$F_c = ma_c = \frac{mv^2}{r}$$

Our definition of force from Chapter 5 is

$$F = k \times \frac{dE}{dr} = \frac{mv^2}{r}$$

and with Energy defined by

$$E = mc^2$$

we can substitute for

$$k\frac{dE}{dr} = \frac{mc^2}{r}$$

to arrive at something that looks like energy going in a circle of radius (r) But what should (r) be?

If we look at a travelling wave, take its wavelength and wrap it around a circle having a radius given by

$$r = \frac{n\lambda}{2\pi}$$

where (λ) (lambda) is the wavelength of the travelling wave, and (n) is any integer from one to infinity.

In theory, any integer number of wavelengths can be wrapped in a circle without destructive interference.

We can also convert a mass into a wavelength and see what the radius is if it is wrapped once around a circle, to get a radius called the reduced Compton wavelength. This is accomplished by using,

$$E = mc^2 = hf = c/\lambda$$

and

$$r = \lambda / 2\pi$$

where (h) is Planck's constant, (f) is frequency, (λ) is the wavelength, and (r) is the reduced Compton wavelength.

When we do this for important particles—the electron, proton, and neutron—the radii we get are as follows:

Particle	Electron	Proton	Neutron	
Mass	9.11E-31	1.67E-27	1.67E-27	kg
Energy	8.19E-14	1.50E-10	1.51E-10	J
Frequency	1.24E+20	2.27E+23	2.27E+23	s-1
Wavelength	2.43E-12	1.32E-15	1.32E-15	m
Radius	3.86E-13	2.10E-16	2.10E-16	m

Table 13.1. The radius of an electron, proton, and neutron

The radius of the proton and neutron derived above are close to the measured values of 0.84 to 0.88E-15 metres, or about 1.5 to 1.57 times larger. Although it is probably a meaningless coincidence, 1.57 radians are equal to 90 degrees.

This exercise was conducted to evaluate whether the equivalent wavelengths of these particles, when wrapped in a circle, had any resemblance to their sizes measured using other techniques such as spectroscopy or nuclear scattering experiments. More can be found by searching "proton radius puzzle" on Wikipedia. There does seem to be a ballpark similarity. Here we also encounter a mysterious ratio.

THE MYSTERIOUS 137

Interestingly, the radius of the electron derived this way is approximately 1/137 the accepted radius of a hydrogen atom from quantum mechanics. Perhaps another meaningless coincidence, but the number is an interesting one in physics.

The number 1/137 is a mysterious number called the fine structure constant, and it appears throughout physics somewhat magically. Richard Feynman (1918–1988), not just a theoretical physicist but a real person, said the fine structure constant is "a magic number that comes to us with no understanding."

The value of 1/137 seems to be related to several fundamental constants,

$$1/137 = \frac{keQ^2}{hc}$$

where (ke) is Coulomb's constant, (Q) is the charge on the electron, (h) is Planck's constant, and (c) is the speed of light.

MATH GOING IN CIRCLES

The next topic may just amount to math going in circles; it is included just for fun but may not signify anything important.

Perhaps it is just coincidence that invoking $(E = mc^2)$ produces the following sequence:

$$k\frac{dE}{dr} = \frac{mv^2}{r} = \frac{mc^2}{r}$$

by substitution of $(v=c)$ for electromagnetic waves. The line of reasoning leads down a mathematical rabbit hole, as now shown:

$$k\frac{dE}{dr} = \frac{mc^2}{r}$$

$$dE = \frac{mc^2}{r}dr = mc^2\left(\frac{1}{r}\right)dr$$

We can integrate energy and the term $(1/r)$ and obtain the following:

$$\int_0^E dE = mc^2 \int_0^r \left(\frac{1}{r}\right)dr$$

$$E = mc^2 = mc^2\left(\ln(1) - \ln(r)\right)$$

However, $ln\,(0)$ is negative infinity, so we run into a fatal error.

But perhaps integrating from 0 to (r) is the problem. If we instead say we will integrate from (r) to 1, the equation reduces to

$$\frac{mc^2}{mc^2} = 0 - \ln(r) = 1$$

We can now solve for the radial distance, (r).

$$r = \exp(-1) = 1/e$$

where (e) is Euler's number, a mathematical constant that is irrational in the mathematical sense: it cannot be expressed as the ratio of two integers. What could this mean, if indeed it means anything?

If we take the mass of the smallest quark, the down quark, and convert it to an equivalent wavelength, then wrap that length around a circle, the ratio of the radius of the circle to the energy in the quark is $(1/e)$, within the uncertainty of the mass of the down quark. However, that is probably just coincidence.

EULER'S E

The number (e) shows up in interesting places, such as Euler's identity:

$$e^{i\pi} + 1 = 0$$

And more relevant to the discussion at hand, Euler's formula:

$$e^{ix} = \cos x + i \sin x$$

Euler's formula describes a circle in the complex plane, as shown in Figure 13.2. In this plane, the x-axis represents the real part of a complex number, and the y-axis represents the imaginary part.

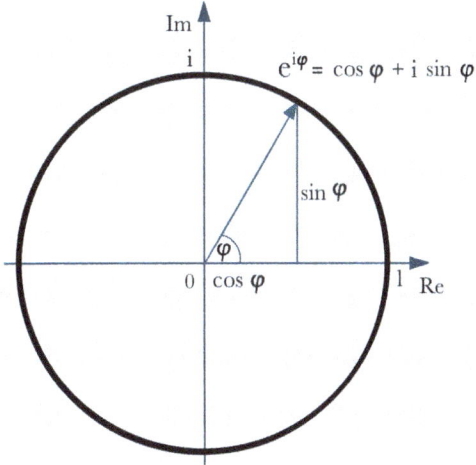

Figure 13.2. The complex plane

Compare this with the space-energy diagrams shown previously . . . and the real part corresponds to 3D space and the imaginary part corresponds to the energy dimension. A mathematical curiosity, not meant to say that the energy dimension is imaginary!

NOW, LET'S C . . .

Why is the equivalence of matter and energy mediated by the factor (c^2), the speed of light squared? The best answer so far is that the conversion factor must have the units of a velocity, metres/second squared, to make the equation work. It can also be shown that the speed of light (c) is the velocity that works when the Lorentz factor is approximated by a Taylor series expansion:

$$\gamma = \frac{1}{\sqrt{1 - \frac{v^2}{c^2}}} \sim 1 + \frac{1}{2}\frac{v^2}{c^2}$$

To make the right-hand side compatible with the total energy of an object with mass including kinetic energy, we must multiply by mass and (c^2), yielding,

$$Total\ E = mc^2 + \frac{1}{2}mv^2$$

SPEAKING OF MATH

There is a lot of strange math and curious numbers in this chapter. They seem to lead to more questions than answers. Odd coincidences that may be meaningless, or useful insights into the space-energy model. Further work is required to determine if there is some deeper meaning to all of it. Fear not, though, for "beyond the scope of this book" is the tried-and-true escape phrase at this point.

The main idea in this chapter is that mass and energy seem to be interchangeable forms of the same thing, possibly both are kinds of electromagnetic waves, either propagating or standing.

In addition, we have seen that while mathematics is the language of the laws of the Universe, sometimes it is hard to tell what it is saying. We can make calculations and develop elaborate theories within the framework of mathematics, but do they always have a physical meaning? That is a good rhetorical question to stop on.

14

FORCES STRONG AND WEAK

"String theory is based on the simple idea that all the four forces of the universe: gravity, the electromagnetic force, and the two nuclear forces, can be viewed as music."

— **Michio Kaku (1947–)**

"Life is so short. I would rather sing one song than interpret the thousand."

— **Jack London (1876–1916)**

THE FOUR FORCES

The quest to unite the four fundamental forces in nature is like the quest for the Holy Grail of physics. The search for a grand unified theory (GUT) that would explain all observations in the world of particle physics with one over-arching framework has compelled scientists since electric and magnetic forces were united by James Clerk Maxwell (1831–1879).

Speculation is that the four forces—gravity, electromagnetism, the strong and the weak nuclear interactions—were all one in the infancy of the Universe, when the energy level was extremely high.

The electromagnetic force and the weak force have been shown to have been united when the temperature (and energy density) of the Universe were much higher, just after the Big Bang. Three physicists, Sheldon Lee Glashow, Steven Weinberg, and Abdus Salam developed the unifying electroweak interaction, which successfully predicts the mass (energy) of force carrying particles called W and Z bosons.

If we accept the contention in this book that any force is a result of an energy gradient the four forces are in a sense unified. A concentration of mass, causing a gradient in the energy background, results in the force of gravity. A concentration of charge presumably has similar electromagnetic effects, producing electrostatic forces. As the Universe seeks energy equilibrium, the phenomena known as entropy, any localized accumulation of mass or charge will give rise to gradients and, consequently, forces tending to remove those gradients.

Suppose the four fundamental forces are related, being responses to concentrations of mass or charge. Gravity operates on the large scale of stars and planets but is only an attractive force and is orders of magnitude weaker than the electrostatic force at small scales. The electrostatic force can be attractive or repulsive, as there are two types of charge, positive and negative. These distinctions between electrostatic forces and gravity may indicate that they cannot be perfectly united, even though they are related. An analogy might be the force of a wave in the ocean and the electrostatic force between polar water molecules attracting each other; both are related to the behaviour of water but operate on completely different size scales.

PARALLEL FORCES

Now we will look at the strong and weak nuclear forces in the context of the electrostatic force and discover interesting parallels. The goal is to show that the strong and weak forces may be due to combinations of electrostatic forces operating on the subatomic scale, where charges are separated by extremely small distances.

Coulomb's law was published in 1785 by the French physicist Charles-Augustin de Coulomb (1736–1806) and was seminal in the development of the theory of electromagnetism. Mathematically, the electrostatic force law appears as

$$F_e = k_e \frac{q_1 q_2}{r^2}$$

The equation bears a striking resemblance to Newton's law of gravity published in 1687:

$$F_g = \frac{G m_1 m_2}{r^2}$$

The two constants (k_e) and (G) have different values, and the charges (q_1) and (q_2) are replaced by two masses (m_1) and (m_2), but the basic structure of the laws is parallel.

These forces act in three-dimensional spheres around the mass or charge, and because the surface area of a sphere is proportional to the square of its radius, the forces at a given radius are inversely proportional to that radius squared. Essentially, the force being spread over a larger spherical surface area that causes it to weaken this way as the distance increases. Figure 14.1 is a diagram showing the concept.

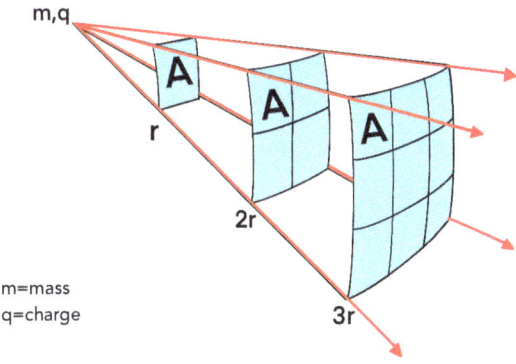

Figure 14.1. Inverse square law

SPECULATION ABOUT ELECTROMAGNETIC WAVES

The three fundamental properties that completely describe a black hole or an elementary particle are mass (energy), electric charge, and spin. The two force laws above are dealing with two out of three of those properties—mass and charge—and that puts an intriguing "spin" on things.

Within the context of the ideas presented in this book, I would go as far as to speculate that electromagnetic waves are fundamentally related to a charge spinning while propagating along one dimension within four-dimensional space-energy. The only problem is photons are not charged unless they are a positive and a negative locked together. Photons do have a spin number in quantum theory. Sinusoidally, varying electric and orthogonal magnetic fields are integral to the picture of these waves, as seen in Figure 14.2.

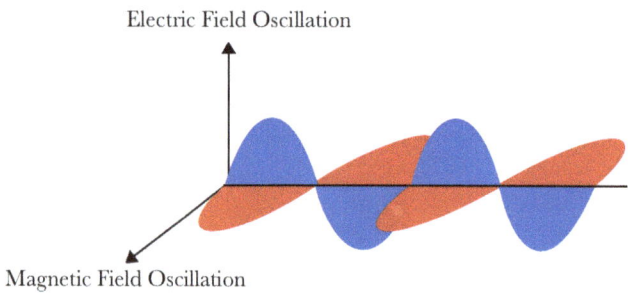

Figure 14.2. Electromagnetic wave

The pattern is much like the wave pattern generated when something spinning is moved, but the electric and magnetic effects occur simultaneously in two orthogonal dimensions. A curious question: Would a spinning object existing in a fourth dimension create what we consider to be an electromagnetic wave by propagating along one dimension while dipping, like the oars of a rowboat, into the two remaining orthogonal dimensions of space?

The energy in these fields disappears and reappears in three-dimensional space, at least in our representation in the figure; so where does it go? The logical answer, in the context of the space-energy hypothesis is into the fourth dimension of energy, although that is admittedly speculation.

Conventional wisdom is that particles such as photons are not an object spinning, but they are said to possess a characteristic called spin. In quantum mechanics, elementary particles have an intrinsic angular momentum. We do not know what charge is exactly either, but we do know it comes in positive and negative versions.

Suppose for the sake of simplifying a complex situation, that whatever it is that is spinning carries the energy by virtue of its spin frequency, and that it is the particle side of the wave-particle duality that will be discussed in a subsequent chapter.

SPLITTING THE BILL—SEPARATION OF CHARGES

Let us now draw a parallel based on the new paradigm of gravity. We will hypothesize that discrete charges, like discrete masses, cause depressions in the overall electric charge background and, therefore, produce gradients within it. On average, at the scale humans are familiar with, the charge background is neutral. When charges concentrate on these scales, they quickly neutralize, as, for example, when lightning resolves an unbalanced charge between clouds and the surface of the Earth.

There is a net charge balance on human and cosmic scales but not at sub-atomic scales, where charges are separated by tiny distances. One might compare this to the ocean, replete with waves having peaks and troughs at human-size scales, but with an average level that remains consistent on the planetary scale, excluding global warming effects.

On the subatomic scale, there are concentrations of charge, down to the level of the fundamental particles. Protons, neutrons, and electrons that make up atoms carry specific quantities of charge. Protons and electrons, have positive and negative charges respectively, of 1.60×10^{-19} Coulombs. A Coulomb is a unit of electric charge defined as one ampere-second of electron current flow: equivalent to about 6.24×10^{18} electrons.

The fundamental charges of the proton and neutron are further divided into fractional values by the theory of quarks proposed by Murray Gell-Mann (1929-2019) and George Zweig (1937-) in 1964. Confirmation followed with the physical discovery of quarks in 1968 at the Stanford Linear Accelerator Center. The topic of quarks, the six types found to exist, will not be explored in detail in this book, but information can be found in various references.

For our purposes, it is sufficient to know that electrons are fundamental particles with a charge of -1, and protons are made up of three quarks with a net charge of +1, while neutrons contain three quarks with a net charge of zero. Figure 14.3 shows the accepted internal structure of protons and neutrons.

Proton   Neutron

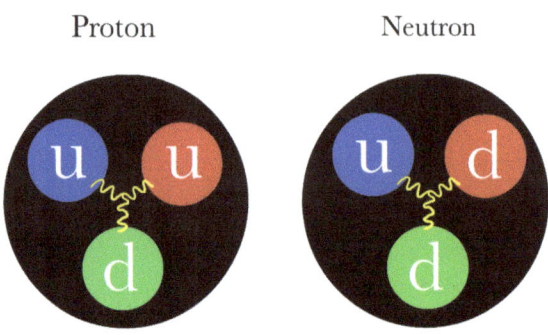

Figure 14.3. Quarks in the proton and neutron

The quarks that make up protons and neutrons are called up and down, and they contain fractional charges of +2/3 and -1/3, respectively. The proton contains +2/3, +2/3, and -1/3, for a total of +1 charge. The neutron contains +2/3, -1/3, and -1/3, for a total charge of zero. Elementary particles called gluons are said to mediate the forces between the quarks. Gluons have mass (energy) but no charge, so they are free . . . a subatomic scale joke there, my apologies.

The foregoing lays the groundwork for a discussion of the strong and weak nuclear forces. I hope to demonstrate they are electromagnetic in nature. That would unite them with the electromagnetic force, a significant part of the Holy Grail of physics. Crucial to this is the idea that charges are separated within the proton and neutron.

THE WEAK FORCE

The weak force is involved in nuclear fusion and fission, and essentially transforms a neutron into a proton plus an electron and excess energy that appears as an electron antineutrino. One way of looking at the process is a neutrino with no net charge closely approaches the neutron and donates a positive charge in the form of a W boson, leaving it as a negatively charged electron.

Rather than going into the details of the Standard Model of Particle Physics, we will take the simplified view that all particles are simply packages carrying different quantities of mass, charge, and spin. Mass has an equivalent energy; therefore, particles are equivalently discrete packages of energy, charge, and spin. These fundamental quantities are conserved, so particles that make that possible appear on cue.

Inside the neutron, from a quark perspective, one of the down quarks transforms into an up quark by absorbing the positively charged W boson, and an electron antineutrino is emitted. These reactions conserve the energy, charge, and spin present.

$$n \rightarrow p + e^- + \bar{v}_e \quad \text{(Beta decay, hadron notation)}$$

$$udd \rightarrow uud + e^- + \bar{v}_e \quad \text{(Beta decay, quark notation)}$$

Neutron decay is quite common, occurring in 886 seconds on average. Protons are far more stable and essentially never decay, or at least are believed to take longer than the life of the Universe to decay, so effectively never.

Perhaps we can consider the proton, being slightly less massive than the neutron, to be a lower energy state than the neutron, so the transformation lowers the energy state. This descending energy transformation is the weak force, in accordance with the previously proposed equation,

$$F = k \, \frac{dE}{dx}$$

Interactions on the subatomic level are conventionally viewed as being mediated by the exchange of particles. In the case of neutron decay, the exchange transfers a positive charge to the neutron and releases energy through the mediating W boson and electron antineutrino. However, it might be useful to view them as quantized exchanges of energy, charge, and spin.

The total amount of charge in a neutron is higher than in a proton, and the two have similar physical dimensions. Neglecting polarity, the proton contains a total charge of 5/3 (two up quarks with positive 2/3 each, and a down quark with a negative 1/3), while a neutron, if we consider it to be the combination of a proton and an electron, contains a total charge of 8/3 (a proton and a negative 3/3 or 1 for the electron). Thus, when a neutron decays, two products with less absolute charge concentration emerge. The process results in a concentration of 8/3 absolute value charge dispersing into a +5/3 proton and -3/3 electron. Protons being a stable configuration of three quarks do not decay. Electrons are the least massive charge carrying fundamental particle and cannot decay into anything smaller. The decay of a neutral particle into two oppositely charged particles seems counter-intuitive since opposite charges attract, but it happens very frequently.

On the cosmic scale, the opposite occurs when a neutron star is formed at the end of the life of a massive star. Stars maintain a balance of inward-acting gravitational compaction force and outward-acting radiation pressure. When the nuclear reactions producing radiant energy weaken, gravity wins out and compresses the constituent protons and electrons into neutrons. Extreme gravitational forces are involved. The collapse either stops when the neutrons cannot be further forced together or, in more massive stars, continues until a black hole is formed.

Taking the view that the forces that cause any events to occur are energy differences, we visualize those differences manifesting as the particles that are believed to be exchanged in the Standard Model, conserving the three fundamental qualities, mass (energy), charge, and spin. As mentioned previously, on the cosmic scale, black holes are also completely characterized by their mass, charge, and spin.

Returning to the weak force, we assume charge is separated within protons, as it is carried by three distinct quarks. If charge can also be separated within the neutrino that triggers neutron decay, the neutrino being equivalent to a positive W boson and a negative electron, then geometry makes it possible for a net electrostatic force to act at short distances between particles and cause neutron to decay. The mechanism for this electromagnetic interaction will be demonstrated in the following discussion of the strong force.

THE STRONG FORCE

The strong interaction or strong force keeps quarks together inside protons and neutrons and keeps protons and neutrons together in atomic nuclei. Protons, being positively charged, repel other positively charged protons, raising the question of how atomic nuclei can remain in one piece? The mutual repulsion of the protons is said to be exceeded by the attractive strong force, keeping the nucleus intact. However, and of prime significance, neutrons must also be present in any nucleus with more than one proton. Neutrons provide a possible mechanism for an electrostatic basis for the strong force.

In all but the simplest element, hydrogen, with its nucleus consisting of a single proton, there are multiple protons married with neutrons. The number of protons gives each element in the periodic table its characteristic atomic number and specific physical attributes: carbon, gold, uranium, etc.

Nuclei with more than one proton always contain neutrons, and the total number of protons and neutrons gives the mass number of the element. Different isotopes of an element result from different neutron counts in the nucleus. For example, carbon has six protons, but its isotopes carbon 12 and carbon 14 contain six and eight neutrons, respectively.

Apparently, the presence of neutrons allows the protons to be bound together by the strong force. However, we can compare the strength of the strong force to the electrostatic force and find that the distance between charges can be selected to give the electrostatic force a strength comparable to the strong force.

The other defining characteristic of the strong force and the weak force is that they only act over very short distances, unlike electric and gravitational forces that in theory act over infinite distances. The strength of electric and gravitational forces decreases of proportionally to $(1/r^2)$, where (r) is the separation distance between masses or charges.

Now let us look at deuterium, the isotope of hydrogen with one proton and one neutron, whose nucleus is called a deuteron. Figure 14.4 depicts the deuteron and the charges contained in the components. The proton and neutron are roughly the same size, having a radius of $0.84 \ 10^{-15}$ metres or 0.84 femtometres. We will simplify the situation and say the proton has a positive charge of +1, and the neutron has a positive and a negative charge that cancel out but can be separated inside the neutron. The logical alignment of these charges is shown below.

Proton   Neutron

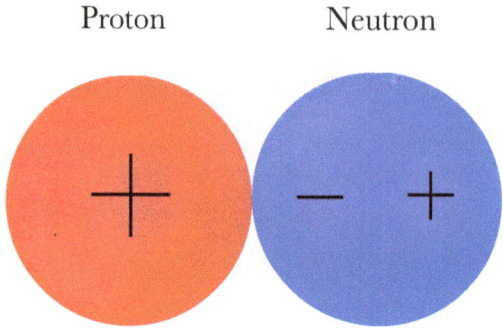

Figure 14.4 Deuteron

When we calculate the attractive force of the proton and the negative charge in the neutron at an arbitrary distance of 1 femtometre (just slightly more than the radius of a proton), we get a force of 231 Newtons.

Then if we calculate the repulsive force between the proton and the positive side of the neutron (assuming it is one proton radius further away), we get a much lower force, only 112 Newtons.

The attractive force is stronger than the repulsive force, so the two will stay bound together. The net attractive force at these distances is 119 Newtons. Note that we are just calculating electric forces due to positive and negative charge, no special strong force is invoked.

The calculation can be repeated over a range of distances, and the attractive and repulsive forces added together to get a net force. The net force is only significant over a short range and becomes roughly eight orders of magnitude weaker at a separation distance of 100 femtometres. At 1 metre of separation, the net force is effectively zero. Table 14.1 contains the results at various distances, beginning with the smallest distance considered possible, the Planck length at 10^{-35} metre, and ending at 1 metre. The net force is displayed graphically in Figure 14.5.

Positive to Negative Charge Separation Distance (metres)	Attractive Force (Newtons)	Positive to Positive Charge Separation Distance - One proton radius greater (metres)	Repulsive Force (Newtons)	Net Force (Newtons)	Comparable Force Magnitude
1.00E-35	2.31E+42	4.38E-16	1.20E+03	2.31E+42	
1.00E-19	2.31E+10	4.39E-16	1.20E+03	2.31E+10	
1.00E-18	2.31E+08	4.39E-16	1.19E+03	2.31E+08	
1.00E-17	2.31E+06	4.48E-16	1.15E+03	2.31E+06	
1.00E-16	2.31E+04	5.38E-16	7.96E+02	2.23E+04	Strong Force
1.00E-15	2.31E+02	1.44E-15	1.12E+02	1.19E+02	Electric Force
1.00E-14	2.31E+00	1.04E-14	2.12E+00	1.90E-01	
1.00E-13	2.31E-02	1.00E-13	2.29E-02	2.01E-04	Weak Force
1.00E-12	2.31E-04	1.00E-12	2.31E-04	2.02E-07	
1.00E-11	2.31E-06	1.00E-11	2.31E-06	2.02E-10	
1.00E-10	2.31E-08	1.00E-10	2.31E-08	2.02E-13	

Table 14.1. Net electrostatic forces

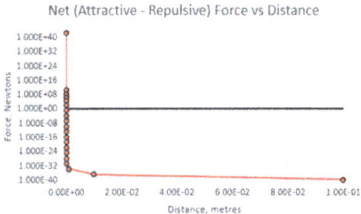

Figure 14.5. Net force vs distance

We can compare the relative strengths of the weak and strong force to the electrostatic force and find charge separation distances where they match the observed values. Establishing a basis for this comparison, at 1 femtometre, the strong force is said to be about 100 times as strong as electromagnetism and a million times stronger than the weak force. The relative magnitudes are highlighted Table 14.1 in the column titled "Comparable Force Magnitude." Reducing the separation of the charges by a factor of 10 causes the electromagnetic force to equal the strong force in magnitude. Increasing the separation by a factor of 100 matches the weak force.

This raises the question: Do we completely understand the location and size of whatever it is that carries electric charge? We can say that if Coulomb's law holds true at the scale of subatomic particles, it is quite possible to replicate the magnitude of the strong and weak forces with the electric force. If various charges and separation distances are allowed, it may be possible to account for many phenomena with electrostatic forces and the energy gradients they imply.

Variations in these calculations are certainly possible, with quarks possessing fractional charges, different nuclear geometries, and various separation distances. Quarks are thought to have a radius in the range of 10^{-19} metres, one ten thousandth of the radius of a proton, and the forces involved become one hundred million times greater at such small proximity.

Exactly how the charge is distributed inside protons and neutrons is not known. Likewise, the exact arrangement of the quarks, if they do have precise locations, is a mystery. Larger atomic nuclei have complex arrangements as conceptualized in Figure 14.6, with many charges imbedded within.

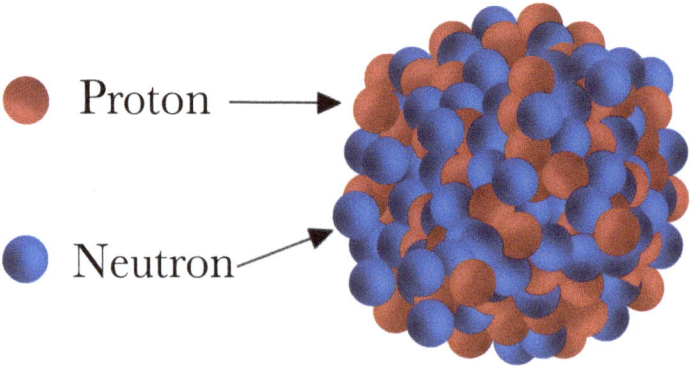

Figure 14.6. Uranium nucleus

Much additional calculation and speculation could be undertaken, but for our purposes, this exercise is only intended to demonstrate that the weak and strong forces can be duplicated in magnitude by adjusting the distance between two charges. In addition, and of major significance, a force that decreases with distance on the atomic scale can be created by combining the infinite acting interactions of opposite charges at varying distances.

Perhaps energy gradients at the subatomic level are not smooth curves but discrete stair steps. Taking a step down produces a particle in some cases such that the total energy and charge and spin are conserved. This would fit nicely with quantum behaviour. If, in all circumstances where conditions permit, a minimum energy configuration is transitioned to by emitting excess energy in particles or electromagnetic waves, it would fit nicely with ever-increasing entropy.

THE HIGGS FIELD

One final point regards the Higgs boson, discovered in 2012 after a 40-year search, using the Large Hadron Collider at CERN. This massive particle is thought to give mass to other particles (those that have mass—not all do) as they interact with the Higgs field. The Higgs field is scalar field that has a non-zero value everywhere in space. As it relates to the space-energy idea in this book, it is suspiciously like the proposed energy background, although there may not be any real connection.

The energy background in the space-energy hypothesis gives inertial mass because mass is simply a localized concentration of energy (a standing wave as opposed to a propagating wave), and the depression that a mass concentration causes in the energy background (its gravity well) must move as the mass moves. This is the mechanism through which the interaction of mass with the energy background occurs. Is this just another way of looking at the Higgs field and the Standard Model? I can only speculate and wonder.

A GUT FEELING

Questions that remain:

- Are the weak and strong interactions electrostatic in nature?

- Can they be regarded as arising from gradients in an electrostatic field?

- Are the electromagnetic, weak, and strong forces produced by charge concentrations, a parallel mechanism to gravity produced by mass concentrations?

If the answers to the above are yes, unification is accomplished, and a grand unified theory is at hand.

QUANTUM MECHANICS **15**

"It seems as though we must use sometimes the one theory and sometimes the other, while at times we may use either. We are faced with a new kind of difficulty. We have two contradictory pictures of reality; separately neither of them fully explains the phenomena of light, but together they do."

— **Albert Einstein (1879–1955), on wave-particle duality.**

QUANTUM MECHANICS

Having explored the large-scale cosmos, it seems appropriate to devote one chapter to quantum mechanics, and to see if the space-energy approach can shed any light on the puzzles and paradoxes of this realm.

A fundamental distinction between classical physics and quantum physics is one of uncertainty. Classical physics provides a means to predict exact future outcomes, whereas quantum physics deals in probabilities. The classical picture is painted with precise measurements of the position, velocity, momentum, and energy content of objects like billiard balls, and the motions of such objects follow predictable trajectories. In the quantum world, there is always an uncertainty associated with position and momentum of particles; knowing one parameter precisely means that the other becomes extremely uncertain, potentially having almost any value. This conundrum could be a consequence of our measurement tools, or an inscrutable quirk in the workings of the world that we can never resolve.

Further distinctions can be made between the classical and quantum perspectives. In classical physics, quantities can vary continuously, but quantum physics permits certain parameters to only vary by discrete, quantized amounts. For example, an electron jumps from one energy level to another in the orbitals of an atom by absorbing or emitting specific amounts of energy. These energy quanta appear as emission or absorption lines of definite frequency on the light spectrum, called Fraunhofer lines after Joseph von Fraunhofer (1787–1826).

Before quantum physics, particles were particles and waves were waves. Now, there is wave-particle duality, and the method of observation impacts the attributes we see. Classical physics is more applicable to describing the behaviour of matter and energy on the scale familiar to humans, and quantum physics deals with the very smallest scale phenomena.

Fortunately, the conservation laws of classical physics also apply in quantum physics. Properties including energy, linear and angular momentum, electric charge, and spin are constant in closed systems. Conservation laws provide a useful bridge between two very different ways of looking at the world.

QUANTUM DICE

Albert Einstein spent many years trying to refute quantum theory despite having contributed to its discovery by describing the photo-electric effect resulting from light quanta or discrete packets of energy. He never considered the probabilistic nature of quantum physics to be the fundamental reality and famously wrote to the quantum theory proponent Max Born (1882–1970), "I, at any rate, am convinced that He is not playing at dice," in reference to God's orchestration of the Universe.

I am taking Einstein's side. Although the discussion presented here is far from a thorough treatment of quantum physics, there are certain common-sense points to be made. Quantum physics is far from common sense, though, so be warned that what follows is indeed uncertain.

I am personally not a big fan of probabilities as an explanation for everything and, in fact, once wrote a paper for the Society of Petroleum Engineers decrying their use in situations where they do not apply (Forth, 1997). Let's use a big fan, anyway, to describe a situation from classical physics that generates a distribution of probabilities.

Imagine, in your own thought experiment lab, an ordinary fan that might be used to cool a room, just an arrangement of rotating paddles that can be spun at various speeds. Now imagine that this fan is located behind a curtain. Imagine we do not know what is behind the curtain but wish to figure it out by shooting pellets through the curtain. Our system allows us to tell if the pellet has hit anything or not as it passes through the curtain and toward the fan, either impacting on a blade or going through unobstructed.

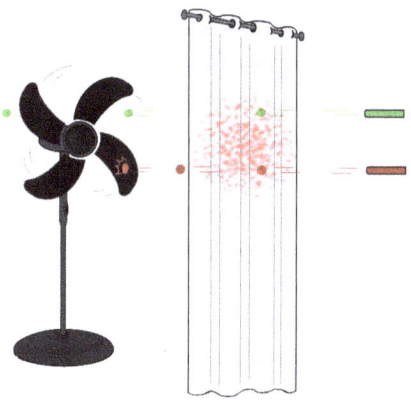

Figure 15.1 Fan behind the curtain

We then fire away, extensively, and all over the curtain, noting the coordinates of each hole in the curtain and whether a hit or a miss occurs. We would find that outside of the radius of the circle described by the turning blades we never get a hit. The probability of a hit in this region after innumerable shots is found to be zero. As we move in toward the centre of the fan, perhaps in some sort of gridded search pattern, we would find that occasionally we would get a hit and occasionally a miss and would be able to calculate probabilities of each outcome from the data. Probability of a hit is defined as the number of hits divided by the total number of shots at a given position on the curtain.

After firing many shots all over the curtain and observing what occurred, we could construct a probability distribution. The appearance would be a circular region with higher probability of hitting an obstacle nearer the centre. This is in line with the quantum result, a distribution of probability of getting a hit, much like the distribution of probability of finding a particle in a certain location. Nevertheless, the reality is a fan spinning behind a curtain, as could be described deterministically by classical physics.

WAVE-PARTICLE DUALITY

Next, we shall consider wave-particle duality. How can something be both a wave and a particle, and be seen as one or the other depending upon how we observe it? Here is a very simple parallel. If you have a guitar or similar stringed instrument, you may want to use it to follow/play along.

To get a sound out of a guitar, you must set up a vibrating wave on the strings, as is being done in Figure 15.2. One way to do that is to pluck a string with your finger. The act of doing this requires energy and transfers energy from your finger to the string, causing it to vibrate at a certain frequency. In this example, we pluck the G string, and it sets up a 196-Hz frequency vibration (assuming the guitar is in standard tuning and modern pitch).

Figure 15.2 Quantum Guitar

There are three ways to observe the experiment. One is by touch. We touch the string with our finger, energizing it or muting it (taking out the vibrational energy) to receive a bit of energy in our finger. Since energy has an equivalent mass, we might observe this as, in effect, a particle hitting our finger, as the string, in fact, does. The second way of observing is with our ear. We hear a pressure wave in the air. A virtuoso musician might be able to identify it as a wave with a certain frequency corresponding to the note G. The third way of observing the situation is by sight, and we immediately conclude that reality is a guitar.

Perhaps in this context, wave-particle duality is not that mysterious after all. The question remains though, is there a reality behind subatomic objects that seem to behave as both waves and particles, a figurative guitar of some sort? I think that Einstein was looking to see the guitar, accepting quantum theory as a way of describing nature in terms of particles and waves, but not believing probabilities are the fundamental underlying reality.

REALITY

The space-energy hypothesis has something to say about the existence of reality, even though we might not have eyes or instruments to see it. There is an indication that a definite reality exists, even though we cannot measure it, at present. Referring to Figure 15.3 in which a two-axis coordinate system is pictured, with the three dimensions of space represented on the horizontal axis and the energy dimension on the vertical axis, let us make an observation.

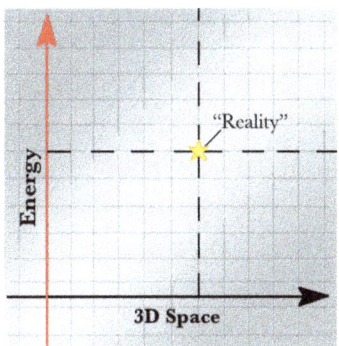

Figure 15.3. Space-energy

Assume that any position on the horizontal axis defines the position of a particle in three-dimensional space, and that any position on the vertical axis defines the momentum (mass x velocity) or corresponding energy (1/2 mass x velocity squared) of the particle.

We know that due to Heisenberg's uncertainty principle, we can only know the position and momentum of a particle to a certain precision. The equation below illustrates the relationship. The combination of momentum and position can not be known more certainly than the quantity defined by Planck's constant divided by (4π).

$$\Delta x \Delta p \geq \frac{h}{4\pi}$$

where (Δx) is uncertainty in position, (Δp) is uncertainty of momentum, (h) is Planck's constant, and (π) is pi.

This implies that if we measure the position precisely, the momentum can be essentially anything, and this corresponds to a vertical line on our graph. On the other hand, if we know the momentum exactly, the position becomes uncertain, corresponding to a horizontal line on the graph.

Note that the lines cross at one point in the graphical representation of space versus energy, and that point is labelled "Reality." Allowing that we could have chosen arbitrarily to measure either position or momentum, it seems that these lines could be determined in principle. And therefore, the coordinates where they cross correspond to a reality that exists but cannot be measured.

Presumably, a given particle has a deterministic position and a momentum, but our measurement techniques introduce the uncertainty. The assertion is that particles have an exact location in space-energy, in the four-dimensional world.

AN EXTRA DIMENSION

The idea of a four-dimensional landscape for reality opens an avenue for investigation of quantum weirdness. Occupants of a two-dimensional world would notice some strange things if they were unaware of a third dimension. Take a tabletop as a two-dimensional flatland world analog, pictured in Figure 15.4.

Touch the surface with two fingertips, then move your hand around. The flat-landers would see two objects, separated but moving in lockstep.

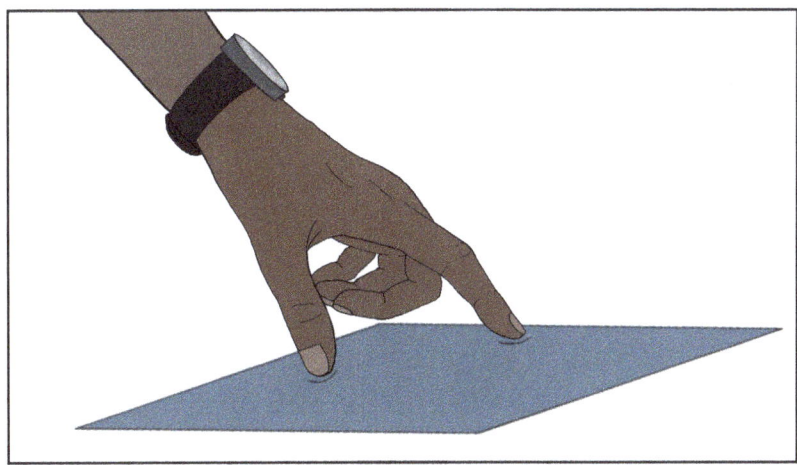

Figure 15.4 Fingers in flatland

In our three-dimensional world, we see "spooky action at a distance." Two particles that are referred to as "entangled" act as if they are connected in some way and seemingly communicate faster than light. This phenomenon is called non-locality, the subject of much wonder and scientific investigation at present.

Does an energy dimension provide a conduit, a place that they are connected, where time does not exist? Recall the projection of two photons travelling along longitudinal strings orthogonal to the equator on a globe from Chapter 4, on the fourth dimension. The photons did meet at the North Pole, yet their projection onto other dimensions could be separated, potentially by any distance. We do not really know how it really works, but scientists are seeking answers.

To summarize, the space-energy model provides a new viewpoint and a new dimension that might help resolve some of the puzzling aspects of quantum mechanics. Professional quantum physicists might be able to see a guitar in all of this, but that remains uncertain.

FERMAT'S PRINCIPLE

16

"Where the statue stood of Newton, with his prism and silent face, the marble index of a mind forever voyaging through strange seas of thought alone."

— William Wordsworth (1770–1850)

"Fortunately for Billy the Splitter his prism sentence was a light one."

MINIMIZING TIME

Imagine a lifeguard in a lifeguard chair located at point A on a beach. She spots a swimmer in difficulty at point B out in the ocean, down the beach some distance. Figure 16.1 is a diagram of the situation. The lifeguard can run down the beach much faster than she can swim in the water. To make the quickest rescue, what path should the lifeguard take—what combination of running on the beach and swimming is fastest?

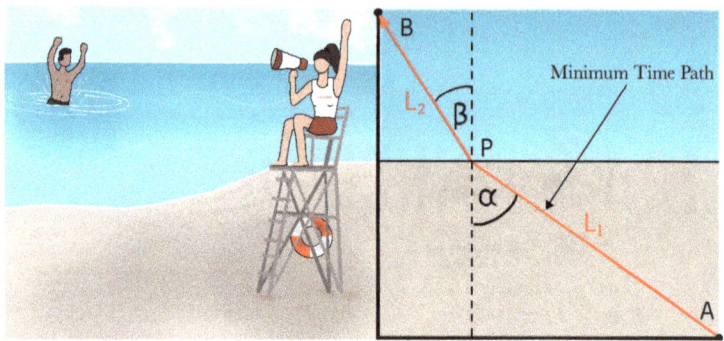

Figure 16.1. Lifeguard in a hurry

Options include heading directly toward the swimmer, or running down the beach to a point where the swimmer is straight out and swimming perpendicular to the beach, or an intermediate path. There is a path that minimizes the total time taken, and time is of the essence.

Pierre de Fermat (1607–1665) developed his eponymous principle, also called the principle of least time, which states that the path taken by a ray of light between two points is the path that can be traversed in the least time. When light passes from air into water or glass, media in which the speed of light changes, it refracts or bends and follows the path taking the least transit time from point A to point B.

In the first chapter on time, we stated that time is just a ratio of distance to velocity ($t = \frac{d}{v}$). Utilizing that approach, we can say,

$$\textit{Total Transit Time Tt} = \frac{\text{distance in air}}{\text{velocity in air}} + \frac{\text{distance in glass}}{\text{velocity in glass}}$$

How does light know the path that will result in the shortest travel time? Clearly even a very bright light does not "know" anything. Could it be that the minimal time path is also the minimal space-energy path, given that nature often seeks the lowest energy solution?

PRISMS AND WINDOWS

Let us speak of prisms and windows after the following light moment: A ray of light walks into a bar of glass . . . and what happens next is sight-splittingly funny! Ahem, right then . . .

If the glass happens to be shaped like a prism, the light splits into colours because white light is composed of a mixture of light of different frequencies that we see as colours. This phenomenon was made famous by Isaac Newton (1643–1727), and later by Pink Floyd's *Dark Side of the Moon* album cover (1973). The illustration in Figure 16.2 below shows light being bent first inside the prism, then again as it leaves the prism. The red light at lower frequency contains less energy and bends less than the higher energy, higher frequency violet light.

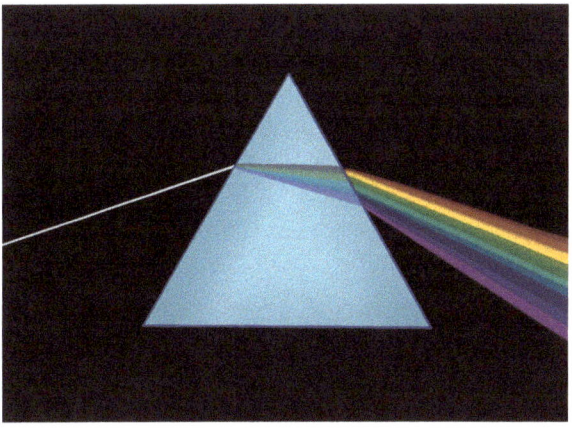

Figure 16.2. Prism splitting light by energy level

However, if the light is going into a slab of window glass, it does not break up into its constituent colours to the same extent but bends as one nearly coherent beam of white light, as shown in Figure 16.3. Fortunate that this happens or everything we see through glass windows would be broken up into a rainbow of coloured images. Light enters the window glass, is deflected, and then emerges from the other side displaced, but parallel to the incident ray.

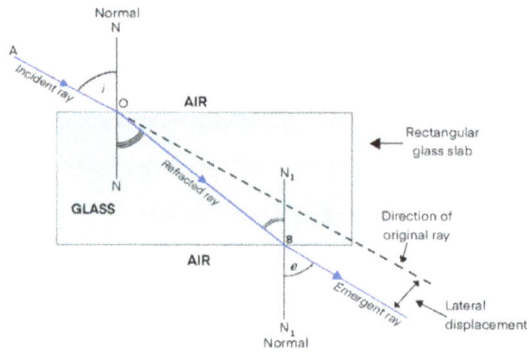

Figure 16.3. A window displacing light

Again, we might ask, how does light "know" it has entered a prism and should separate into colours or into a window and should remain a single beam of white light containing photons of all colours? The only difference, assuming we use the same type of glass for both prism and window, is the geometry of the complete setup. Initial contact with the glass surface is the same in both cases. The light behaves as if it can tell in advance whether it will emerge from a parallel or a non-parallel back side of the piece of glass. This seems impossible.

The Huygens-Fresnel wavefront theory, where light impinges on a surface and then spreads out such that the shorter wavelengths bend more, appears logical; however, the first surface encounter of a window and a prism are indistinguishable—in both instances, there are just waves encountering a new media.

The only way to make sense of what we observe is that the properties of the materials and the complete geometry determine what the light will do, in effect laying out a path for it to follow.

We note that light of different colours follows different paths through the prism, having been bent by different degrees. The energy of a photon of light is related to its frequency by the Planck-Einstein equation:

$$E = hf$$

where (E) is the energy in Joules, (h) is Planck's constant (6.62657 x 10^{-34} J. s), and (f) is frequency (1/s)

The prism is notably different from the window in that there is less width of glass in one direction than in the other. The geometry is not symmetrical. The red light covers less distance in the glass than the violet light. The path that the light follows depends on its frequency, which is directly related to the energy each photon contains. The prism acts as a frequency or energy filter in space.

THROUGH THE PRISM OF SPACE-ENERGY

The idea of an energy background in space, and a uniform mass-energy density throughout the cosmos can perhaps provide an explanation for our observations. Recall in the Universe as a glass of ice water analogy in Chapter 2 that the Universe maintains a constant ice water (mass energy) level, like the constant air pressure everywhere in a balloon. Where there is a mass present, the prism, there is less energy background and a corresponding non-symmetrically shaped divot in the background. For a piece of uniform glass, the energy divot is symmetrical. Where there is less energy density and more mass, the hypothesis suggest light slows down.

The proposed solution to the prism phenomena is that the distance through space-energy for each colour of light traversing the prism must be minimized. To obtain more enlightenment on the idea, suppose we force the blue and red light to leave point A in the air and arrive at point B in the prism. For that to happen, the light must follow different paths through space. The blue light would have to take a longer route through space as illustrated by the blue dashed line in Figure 16.4 to arrive at the same point as the red light.

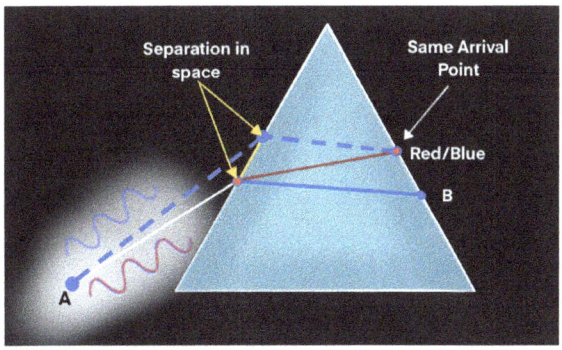

Figure 16.4. Red and blue light – paths from A to B

If the blue light takes a longer path through space, does it follow a shorter path in the energy dimension? All that can be said presently is that the frequency of blue light is higher, and therefore, its wavelength is shorter. The modification of the Minkowski equation in Chapter 4 had wavelength as the term for the energy dimension.

$$x^2 + y^2 + z^2 + (i\lambda)^2 = constant$$

Further work is required, specifically by conducting measurements of the paths of light of specific frequencies from A to B in an optical laboratory. This might confirm a minimal space-energy path numerically.

Let us next open another window on the fourth dimension.

SNELL AND LORENTZ

The equal bending of light through a window such that it emerges on the other side on a parallel but laterally displaced line is also interesting from the perspective of the fourth-dimension idea. At an interface between two media having different refractive indices, light bends according to Snell's law (1621), independently discovered by René Descartes and published in 1637, illustrated below in Figure 16.5,

$$\frac{\sin\emptyset_2}{\sin\emptyset_1} = \frac{v_2}{v_1} = \frac{n_1}{n_2}$$

where ($sin\ \emptyset_1$) and ($sin\ \emptyset_2$) are the angles formed to the normal line to an interface, (v) is velocity in the respective media, and (n) is the refractive index of the material. The refractive index is equal to the ratio of the speed of light in a vacuum to the speed in the material.

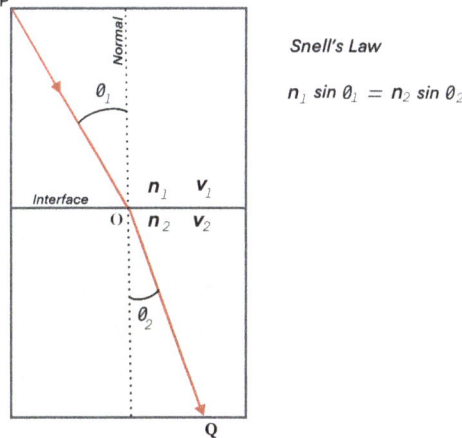

Figure 16.5. Snell's law of refraction

Snell's law corresponds nicely with the Lorentz transformation, as rearranged according to the Pythagorean theorem as was discussed in Chapter 4. The speed of light in a media varies proportionately with its refractive index and is related to the deflection angle through Snell's law. Figure 16.6 is the corresponding space-energy diagram.

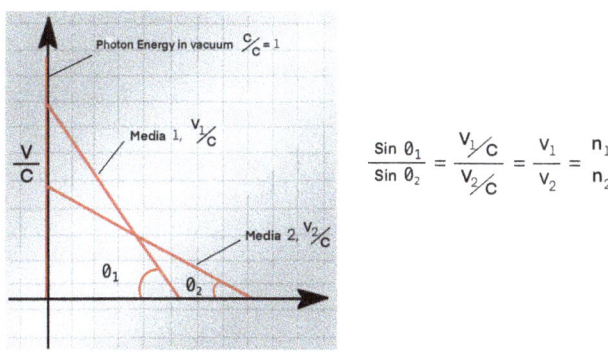

Figure 16.6 Snell's law in space-energy

Light entering a media with a different refractive index changes its speed relative to its speed in a vacuum. The photons retain their energy because their wavelength shortens in proportion to the change in speed, in order to keep their frequencies the same. Frequency is the parameter that relates directly to energy, as noted above.

In Figure 16.6, the red line, denoting the energy of a photon when its speed equals the speed of light in a vacuum, stays constant, even as its relative velocity (v/c) changes. Significantly, this relationship duplicates Snell's law in the context of this representation of space-energy.

We find that the deflection angle into three-dimensional space is given by the Lorentz transformation, another indication of the proposed linkage of an energy dimension and the three dimensions of space. To maintain unity of the Lorentz triangle hypotenuse, a reduction in speed must be accompanied by a displacement in three-dimensional space. When full speed resumes on the other side of the glass, the direction returns to a line parallel to the previous course before the glass was encountered.

There should be no real surprise that two physical principles that have been proven correct are consistent with each other, but this example linking refraction to relativity is illuminating in any colour of light.

LIGHT REFRACTION VERSUS ENERGY CONTENT

If a red and a blue photon are represented on a space-energy diagram, with energy on the vertical axis and denoted by ($E = hf$) instead of (v/c), we can see what happens when they leave a vacuum and enter a media such as glass. The blue photon has more energy and a longer line along the energy axis. Refer to Figure 16.7 in conjunction with this discussion.

The energy content of each photon is given by its frequency. Energy, appearing as a line along the vertical axis, is conserved, so the line length remains the same as they rotate to slide along the space axis, indicating a deflection in space occurs. Both photons are deflected at the same angle as specified by Snell's law and the Lorentz relationship expressed as a Pythagorean triangle.

From this, we can see how a window differs from a prism. The thickness of a window is constant, so both photons travel the same distance through glass and emerge at the same vertical line on the space axis. By contrast, with the prism, the red photon emerges from the glass having traversed a shorter distance than the blue photon when it emerges, as indicated on the horizontal axis in the diagram.

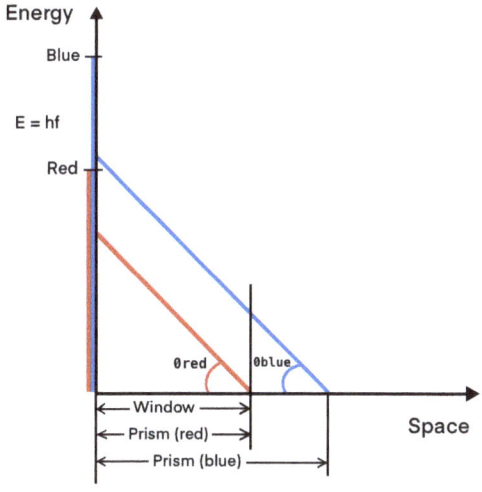

Figure 16.7. Refraction in space-energy

This description fits with what we observe. The initial transition from vacuum to glass media is identical for both situations. However, the blue light and the red light emerge at different positions in space, depending on whether the glass is uniform in thickness or not.

If we look up or down along the vertical energy axis within the plane of the graph, both photons do travel along the same path, but in the space dimensions, the path depends upon the energy or colour of the photon.

Questions remain with the topics in this chapter. These questions do submit to scientific experiment in the laboratory that could be carried out as one test of the space-energy hypothesis. Particularly to determine whether there is any formal relationship between the actual path lengths in space and the energy levels in the photons. Since the forgoing is a thought experiment, the results to date are only "I think so."

UNICYCLE

17

---

*"Evolution, of course, is not something that simply applies
to life here on earth; it applies to the whole Universe."*

— **John Polkinghorne (1930–2021)**

"The wheel is come full circle."

— **William Shakespeare (1564–1616)**

Suppose we suspend disbelief and, for a moment, accept the ideas described up to this point in the book. What then are the implications for the evolution of the Universe? The entire life cycle of the Universe can be accounted for in a new and perhaps surprising way.

EXPANSION OF THE UNIVERSE

To begin at the beginning, we accept that the Big Bang occurred, as it seems to be logical based on our observations of the Universe. The expansion is inferred from the redshift of light coming to us from stars at various distances indicating they are moving away from us. If we reversed the expansion, it would lead to a coming together at a common point of origin.

Having spent considerable effort arguing that time is nothing but a human construct for convenience, a ratio of distance to velocity, I am reluctant to say that we are reversing time to get back to the origin of the Universe. Instead, a course reversal of objects in space is the ticket—what we experience as time began when distances and velocities were initiated as the Big Bang happened. Time, as a ratio of distance to velocity, is always possible if there are distances and velocities available to base it on.

Distance is an important consideration; did the initial distance across the Universe start at zero, in what would be called a singularity? My suspicion is that there never was a zero-dimensional point that contained everything—a suspicion founded on logic that will become clear as we examine a scenario for the evolution of the Universe.

What about velocity, in particular the ultimate velocity, the speed of light? My hypothesis is that the speed of light at the beginning of the expansion of the Universe was much higher than it is today, but it was not infinite. The speed of light is proposed to have been proportional to the square root of the energy density of the Universe, and since the volume is likely to never have been zero (no singularity), the density would have been extremely high but not infinite.

The Universe was forced to begin expanding, because it started as a concentration of energy with an extremely large energy gradient to its surroundings. We might say it was forced to expand by entropy, the inexorable tendency toward energy equilibrium. "Go forth!" said entropy.

Initially, the expansion happened incredibly quickly but always at the prevailing speed of light. Assuming the Universe is spherical, in three dimensions, the volume is proportional to the cube of the radius, and that radius was increasing at the speed of light. The energy density, thus, decreased very rapidly and, with it, the speed of light. The period of incredibly rapid expansion that corresponds to the era of cosmological inflation is an inevitable feature of this model, but no special or mysterious energy is required for it to happen, other than the constant and conserved normal mass-energy in the Universe today.

The birth of the Universe might be likened to a fireworks shell exploding in air. A very rapid initial expansion (due to high energy gradient between the explosive reaction and the surroundings) slowed eventually by air resistance. The cosmos is not expanding into air, so the analogy breaks down, but it is a great visual model. Most of us have seen fireworks, like the illustration in Figure 17.1. Next time you see a display of fireworks, think about the Universe!

The expansion of the Universe in our model is slowed due to the slowing of the speed of light as it travels through a dispersing media, the decreasing electromagnetic energy density associated with a fixed quantity of mass-energy and an expanding volume.

Figure 17.1. Big Bang fireworks

Energy seeks equilibrium. When cool milk is added to hot coffee, they quickly come into thermal equilibrium. The hypothesis is that the entire Universe acts in much the same way, constantly seeking to maintain a constant level of mass-energy intensity on the large scale. The drive toward energy equilibrium (maximum entropy) is what drives the expansion of the Universe forward. Entropy itself is equivalent to the relentless movement toward energy equilibrium, where there are no energy gradients and, thus, no forces.

And so, we have expansion, through to the present stage of the evolution. Science is not sure whether this expansion will slow down due to gravity (a closed Universe), or go on forever (an open Universe), with insufficient gravitational attraction arrest the expansion. Between the two scenarios is the flat Universe; but flat Earthers should not get ideas from this, it is not that! These cases correspond to positively curved space, negatively curved space, or zero curvature space, respectively.

STOPPING THE EXPANSION

The VSL hypothesis of the four-dimensional space-energy Universe utilizes the Schwarzschild solution to Einstein's equations to suggest a mechanism for stopping the expansion of the Universe.

We need to return to the concept of the Schwarzschild radius, discussed earlier, and defined as

$$r_s = \frac{2GM}{c^2}$$

where (r_s) is the event horizon of the Schwarzschild black hole, (G) is the gravitational constant, (M) is the object mass, and (c) is the speed of light.

A given mass (M) when compressed into a sphere of this radius becomes a black hole. The escape velocity becomes the speed of light, and because nothing can go faster than light, nothing can escape.

The current Schwarzschild radius of the entire Universe is estimated to be approximately 23.5 billion light years, using 1.5×10^{53} kg as the amount of ordinary mass it contains, the gravitational constant, and the current speed of light. The radius of the entire Universe is estimated to be about 46 billion light years.

How can the Universe have attained a radius of 46 billion light years when the Universe has only been around 13.8 billion years? The hypothesis of this book is that light speed has been much faster in the past, to the effect that 46 billion light years have been covered in only 13.8 billion years. Mainstream theory explains the difference by allowing that space can expand faster than light.

The key to the anticipated fate of the Universe from the point of view of the space-energy theory is that the speed of light is decreasing. The effect on the Schwarzschild radius is that it increases with decreasing light speed (squared!). Mathematically, the Schwarzschild radius will surpass the radius of the entire Universe and, eventually, turn it into one Universal black hole.

Since, at that juncture, not even light will be able to escape, the expansion of the Universe will cease. Presumably, everything inside will remain as it was because the event horizon will be far beyond what is observable to us.

What follows is speculation but progresses logically. Gravitational collapse sets in. Black holes already exist throughout the Universe; we suspect there are many and that they grow by gravitationally swallowing mass and energy. These black holes inside the Universal black hole should continue to grow and merge. Presumably, the mass of the Universe would continually aggregate into a more compact volume, but not a dimensionless singularity.

We really have no idea what goes on inside a black hole, so the fate of the Universe could be much different, but in this scenario the inside of the Universal black hole would look a lot like our present cosmos, with the exception that everything would be coming together instead of spreading apart. You have to like it when things come together!

I would suggest, though, that since a black hole cannot lose energy, the mass inside cannot possibly collapse into a singularity. But a lot of the mass and energy inside could be sequestered in black holes, whose energy and mass content can be looked at as mass alone. This would lower the energy density of the Universe, slowing the speed of light further.

If the mass came together it might transform into energy of increasing density. The transformation would be more than just molten rock vaporizing, it would be of the ($E = mc^2$) kind, leaving behind mostly pure energy once more. The speed of light within the Universe would increase as a result. The Schwarzschild radius would then contract in what might be referred to as a Big Crunch.

Figure 17.2 illustrates the sequence of events for one cycle of the Universe. Logically, it repeats this cycle indefinitely.

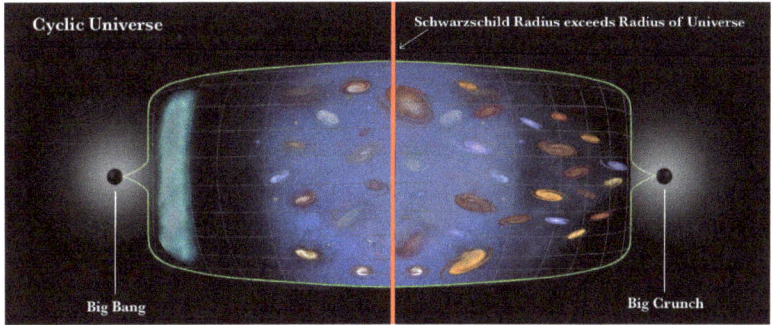

Figure 17.2. The cyclic Universe

A test of the idea now precipitates. In the post-Big Bang era, cooling occurs, and mass condenses out of the energy inferno. Combine this with an expanding Schwarzschild radius, and black holes might begin to form earlier than might be expected to result from the subsequent gravitational collapse of cooled matter alone.

If we find that black holes are evident in the anthropomorphic childhood of the Universe, this idea could be the explanation. Areas of mass lassoed into black holes by loops of expanding Schwarzschild radius, stabilizing with the speed of light as its decline became less precipitous. As mentioned previously, quasars near the edge of the observable Universe may be powered by extremely massive black holes that formed relatively early on.

In addition, the steeper energy gradients in an early post-Big Bang Universe might aggregate mass more quickly that would be possible today. The result would be the earlier appearance of stars and galaxies than we can account for with conventional theories. Indeed, recent JWST observations are finding large galaxies that appear to have come into existence much earlier than expected.

The fate of the Universe, according to the logical development of the space-energy idea, is a cyclical progression, from Big Bang to Big Black Hole to Big Crunch, repeating indefinitely. Each cycle would see the re-emergence of time, assuming it contained beings like us to invent the concept.

PIONEER ANOMALY 18

THE PIONEER ANOMALY
AND LIGHT SPEED IN A GRAVITY WELL

The testing of a scientific hypothesis can take considerable distance/velocity (i.e., time). However, there is a test for this VSL hypothesis that can be conducted immediately. Fortunately, the data needed to conduct this test has been collected and examined in detail by the National Aeronautics and Space Administration (NASA).

Back in 1972/73, NASA launched the Pioneer 10 and 11 spacecraft with the goal of close-up observation of the planet Jupiter, and the planets Jupiter/Saturn, respectively. The craft were the first two objects to achieve escape velocity from the solar system and are now outside the orbit of Neptune, on their way to interstellar space. Due to the unprecedented measurement accuracy of the trajectories of the vehicles an anomalous acceleration toward the Sun was discovered that is not in accord with the laws of gravitation. This raised concerns that there might be something wrong with our understanding of the physics of gravity and launched an intense effort to find an explanation.

A solution was proposed in the 2012 article "Support for the Thermal Origin of the Pioneer Anomaly," by Slava Turyshev et al., which proposed a solution to this minute observed acceleration toward the Sun of 8.74 +/- 1.33 x 10^{-10} m/s. The cause is attributed to anisotropic radiation pressure, radiation of more heat from the craft in a direction that would produce the small sunward thrust needed to explain the slowing down of the spacecraft. The heat is produced by onboard radioisotope thermoelectric generators (RTG). The radiation must be emitted in a way that produces an unbalanced force for this idea to work. Detailed finite element thermal modelling was conducted and concluded that the cause had been identified. Case closed.

CASE RECONSIDERED

Nevertheless, let us consider the problem from the space-energy perspective. Each mass in the solar system has a gravity well, proposed to be a depletion in the energy background as a result of the accumulation of energy into the mass. The energy density of the background can be precisely calculated because we can determine the total potential and kinetic energy needed to place an object in orbit at any distance from a mass. This maps out the gravity well, and the energy density at any distance.

Since the speed of light is proposed in this book to be related to the energy density, we can mathematically calculate the difference of the speed of light on Earth and at the edge of the solar system. The energy background being more intense further from the mass in question means a higher energy density further away and, thus, a higher speed of light further away.

If the speed of light is faster on average through the intervening space than we determine it to be based on measurements made on Earth, the signals from the Pioneer spacecraft would give the impression that they were closer than they really are and that the vehicles must have slowed down.

To elaborate on this idea, imagine your friend has a car that can only go 100 kilometres per hour. Your friend is travelling from 100 km away, so you expect he will arrive in exactly one hour. Instead, he arrives in 42 minutes, much to your surprise! Well, he must not have started 100 km away, he must have started only 70 km away. Unless, of course, the car was really going 143 km/hr. There are two possible explanations, both potentially valid.

So, let us do the math for the Pioneer Anomaly and see what happens.

Figure 17.1 shows the acceleration toward the Sun versus the heliocentric distance in astronomical units (Earth's orbit radius = 1 AU). For simplicity, we assume both Pioneers 10 and 11 behave similarly, and we will take the value of the unexplained Sunward acceleration to be 8.74 x 10^{-10} m/s^2.

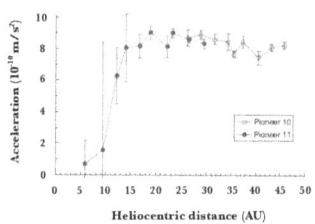

Figure 18.1. Pioneer acceleration vs distance

Using the equations of motion, we can determine what the distance change would be from this acceleration.

We have,

$$S = S_o + v \times t + \frac{1}{2} a \times t^2$$

where (S) is distance, (S_0) is initial distance, (v) is velocity, (a) is acceleration, and (t) is time.

We are only interested in the acceleration term for this exercise. We will consider the time for the spacecraft to reach 30 AU based on Pioneer 11 launch on April 5, 1973, to last good data transmission on November 24, 1995—a span of 8,268 days or 714,355,200 seconds.

The distance impact of the anomalous acceleration, if it was in effect for the entire life of the probes, is therefore estimated to be a distance of,

$$S = \frac{1}{2} \times 8.74 \times 10^{-10} \times (714{,}355{,}200)^2 = 2.233 \times 10^8 \text{ m}$$

with an uncertainty of +/- 3.394 x 10^7 m or 15.2%

Since they launched from Earth at 1 AU from the Sun, an incremental 29 AU is 4.88×10^{12} m, the distance discrepancy as a fraction of the total distance is,

$$2.233 \times 10^8 / 4.338 \times 10^{12} = 5.147 \times 10^{-5} \sim 0.00515\%$$

with an uncertainty of + /- 0.00078%.

Next, we must determine the difference in the speed of light that would be expected from the different background energy densities on the Earth's surface and at the edge of the solar system. The energy density at the Earth's surface is important because that is where we measure the speed of light. We must superimpose the effect Earth's gravity well on the Sun's gravity well by adding the mass of the Earth to that of the Sun, as an approximation. We will assume the same mass density in space at both locations to simplify things, and because signals from the Pioneer spacecraft only spend a tiny fraction of a second in the atmosphere (about 250 microseconds).

The relative speed of light differential will be given by

$$\frac{c_{ess}}{c_{se}} = \sqrt{\frac{E_{ess}/m}{E_{se}/m}} = \sqrt{\frac{E_{ess}}{E_{se}}}$$

the variables being the speed of light (c) at the edge of the solar system (ess) or surface of Earth (se). The energy (E) is the background energy level, as given by orbital energy calculations: the potential and kinetic energy required to maintain an orbit. The mass (m) cancels out. The volume we use to measure the energy density is assumed the same at both locations. The edge of the solar system is taken to be at 30 AU, and the comparison is made to the accepted speed of light in a vacuum on the surface of the Earth, where most modern measurements of light speed have occurred.

The necessary energies are easily calculated, as they are the sum of the potential energy and the kinetic energy necessary to stay in orbit. Recall, no energy gradient means no force, thus no force of gravity.

For the potential energy, we will calculate the maximum value as that required to remove a one-kilogram mass from the Schwarzschild radius to infinity. We must calculate these for the combined mass of the Sun and the Earth.

This approximation is justified by the fact that the Earth orbits the Sun and is on average close to the same distance from the probes as the Sun.

Using

$$K_e = \frac{1}{2}m(v_o)^2 = \frac{1}{2}m\left(\frac{GM}{r}\right)$$

where (v_o) is the orbital velocity, $\left(\frac{GM}{r}\right)$

$$P_{e\,maximum} - P_e = \frac{GMmc^2}{2GM} - \frac{GMm}{r}$$

$$= \frac{mc^2}{2} - \frac{GMm}{r}$$

The total energy is then,

$$T_e = K_e + P_{e\,maximum} - P_e$$

We determine the total energy at the surface of the Earth and at the edge of the solar system, and then take the square root to get the corresponding speed of light difference. The ratio of the change is then calculated for comparison with the distance discrepancy estimated using the acceleration. The result is shown in Figure 18.2.

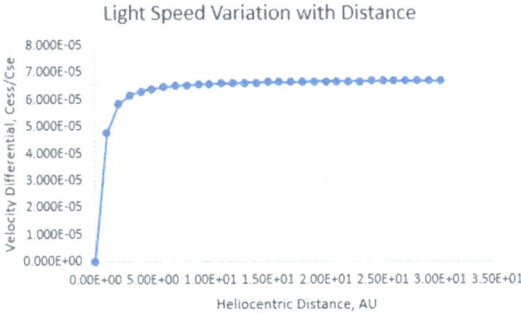

Figure 18.2. Variation in light speed due to energy density

When the calculations are undertaken, the result gets us admission into the ballpark. The discrepancy in distance (5.174×10^{-5}) and in speed of light (6.49×10^{-5}) are within a factor of 1.25 times. The answer is close but not exact.

SPEED OF LIGHT IN A GRAVITY WELL

In the process of investigating the Pioneer Anomaly, an equation was developed that relates the speed of light within a gravity well to the speed of light far from any mass.

First, we derive the maximum potential energy (U) to lift something from the Schwarzschild radius (r_s) to infinity,

$$U = \frac{GMm}{r}, with\ r = r_s = \frac{2GM}{c^2}$$

By substitution, results in

$$U = \frac{GMm}{2GM/c^2} = \frac{1}{2}mc^2$$

The total energy is

$$E = \sqrt{Maximum\ Potential\ Energy - \frac{GMm}{r} + Kinetic\ Energy}$$

$$E = \sqrt{\frac{1}{2}mc^2 - \frac{GMm}{r} + \frac{1}{2}mc^2}$$

Note that strictly speaking kinetic energy at the speed of light is not $\frac{1}{2}mc^2$, but allowing it to be so results in something extremely interesting.

Recall that the speed of light is given by

$$c = \sqrt{\frac{E}{m}}$$

By substitution, we can calculate the speed of light in a gravity well:

$$c_{gw} = \sqrt{\frac{\frac{1}{2}mc^2 - \frac{GMm}{r} + \frac{1}{2}mc^2}{m}} = \sqrt{c^2 - \frac{GM}{r}}$$

Where cgw is speed of light in a gravity well as related to in free space away from all mass.

The result it gives in the case of the Pioneer Anomaly is a 3.695 x 10^{-5} discrepancy in light speed compared to the 5.174 x10^{-5} difference in distance. Now the answer is low by a factor of 0.71. These two methodologies have the observations surrounded. The result is shown in Figure 18.3.

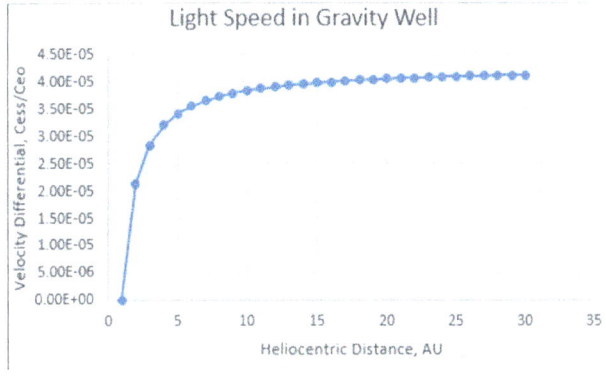

Figure 18.3. Light speed in gravity well

The data show the acceleration was not constant over the entire voyage of the Pioneers. They also did not travel radially away from the Sun but followed complex trajectories. The profile of the acceleration versus distance bears a striking resemblance to the speed variation calculated by the two methods discussed above. Adjusting for this reduces the distance discrepancy to roughly 4 x 10^{-5}, and with the uncertainty in the data, it could match the gravity well calculation.

The above equation for speed of light in a gravity well has implications for black holes and solves a confusing issue. The escape velocity of a black hole is equal to the speed of light. Escape velocity means something can leave a gravitational well and keep going, never to be attracted back. However, objects traveling above orbital velocity but below the escape velocity can still take a trajectory that lets them travel a long way from the primary mass without escaping. The orbital velocity is the speed required to stay in a circular orbit. Velocities intermediate between orbital and escape values are entirely possible.

For a satellite of the Sun, a velocity in excess of orbital would allow a satellite to enter an elliptical orbit, like that of a comet, and to travel far out, possibly encounter another satellite such as Earth and be seen. For light travelling at the normal speed of light in a vacuum around a black hole, the implication is that the light could get away and be seen, at least from just outside the event horizon. In other words, black holes might be visible, as light that cannot technically escape might still be able to go beyond just the orbiting near or falling into the hole.

Plugging the Schwarzschild radius into the equation developed previously, we find that the speed of light slows to $c/\sqrt{2}$, coincidentally the orbital velocity for the black hole.

$$c_{gw} = \sqrt{c^2 - \frac{GM}{2GM/c^2}} = \sqrt{c^2 - \frac{c^2}{2}} = \frac{1}{\sqrt{2}}c$$

We find, by these mathematical manipulations, that (c_{gw}) at the Schwarzschild radius or event horizon is equal to the orbital velocity of the black hole, rather than the escape velocity. The relationship between the two is,

$$v_o = v_e/\sqrt{2}$$

This means light can orbit at the event horizon or fall into the black hole, but it cannot move away from it, any more than a satellite orbiting Earth can move to a higher orbit without energy being applied to it. The ability of the black hole to trap energy and matter is secured absolutely if the speed of light slows in this manner in the gravity well the black hole produces.

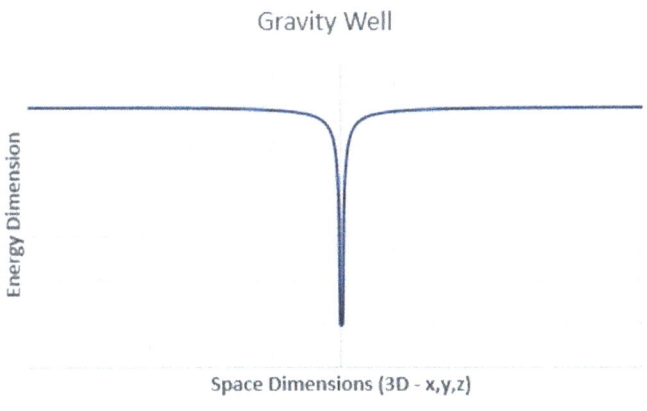

Figure 18.4. Gravity well

The depiction of a gravity well in a cross section through four-dimensional space-energy is shown in Figure 18.4.

The Pioneer data is not generated by the spacecraft moving radially away from the Sun but comes from the actual trajectories, courses that curled through the solar system. The influence of the gravitation of other planets may also cause variations from the simple model used here. A more in depth look at the data, including how much time the spacecraft spent at various distances from the Sun and the effects of gravitational slingshots around Jupiter and Saturn might produce answers closer to the observations.

GRAVITATIONAL POTENTIAL ENERGY–TIME EFFECTS

We do know that time slows down if we lift an atomic clock a small distance within the Earth's gravitational well; the effect has been observed. The space-energy hypothesis predicts the speed of light should increase as the result of a miniscule increase in the energy density of the background as the distance from the mass of the Earth is increased.

The faster carrier speed would translate to in longer wavelengths. If one is counting wave cycles as a measure of time, as atomic clocks count the vibrations of a Cesium atom, longer wavelengths produce a slower count and an apparent slowing of time.

This same phenomenon would produce gravitational redshift and blueshift in light as well. The incremental energy light gains as it falls into a gravity well is reflected in a shorter wavelength, blue shifted light, and the energy lost as it climbs out of the gravity well lengthens the wavelength or redshifts the frequency. These wavelength shifts would also occur if the speed of light were changing with the gravitational potential or background energy density.

The speculations from the space-energy hypothesis are subject to mathematical and experimental testing, as all scientific hypotheses should be.

GOING FORTH

19

"I believe in evidence. I believe in observation, measurement, and reasoning, confirmed by independent observers. I'll believe anything, no matter how wild and ridiculous, if there is evidence for it. The wilder and more ridiculous something is, however, the firmer and more solid the evidence will have to be."

— Isaac Asimov (1920–1992)

"The distinction between truth and rubbish is repeatable experimental proof."

— Aronymous.

GOING FORTH—FURTHER WORK

To generate a collection of ideas is one thing, but if they cannot be tested it may be a pointless endeavour from a scientific perspective. Much of the content presented in these pages is based on intuition and imagination, but as Einstein said, "Imagination is more important than knowledge." Still, the ideas outlined in this book must rightly include real-world tests to determine their veracity.

Thought experiments might seem rational and logical, but that is insufficient. Mathematical arguments can be checked to verify that they do not violate the principles of mathematics—more convincing, but still not sufficient. Physics, by definition, is the branch of science concerned with the nature and properties of matter and energy. Physical confirmation by repeatable experiment in the real world is therefore essential for the preservation of truth.

Many of the ideas in the book are derived in a straightforward way from accepted mathematical equations. Barring calculation errors, the results are difficult to dispute, but real-world proof is still required because unwarranted assumptions can lead down the mathematical rabbit hole.

Several of the concepts in the text will, of necessity, need to be proven by negative outcomes. Scientific investigation may disprove or fail to verify certain contemporary mainstream ideas. The concept of dark matter is a good example. Alternate explanations for the observations that lead to the assumption of dark matter are available. An abundance of ordinary matter that has not been accounted for to date, including less detectable black holes, dim stars, planets, and dust, might be discovered but are all composed of the ordinary matter we presently recognize. No unusual matter, with the ascribed properties of dark matter, has been found to date. The search for dark matter is in progress, and in that sense, the idea that it does not exist is already being evaluated.

On the positive side, there are things that can be done proactively to prove/disprove many of the innovative ideas from the space-energy hypothesis. Since a variable speed of light is postulated, measuring the speed of light with extreme precision over the life of the Universe would be very germane.

At this stage in the evolution of the Universe, the variance expected in the speed of light is exceedingly small, on the order of 0.000015 metres per second per year. The precision in measuring the speed of light over the period since humanity began measuring it in 1676 is likely not sufficient. That is not to say that the measurement precision we have today is not sufficient, but we need precise measurements over a significant interval to pin down any change within the margins of measurement error.

The early Universe was much higher in energy density if the Big Bang is the correct explanation of its evolution. Could we artificially create a local high energy density environment, and see if light moves faster through it? That light slows down in the presence of mass is already well known. This is why the speed of light is always specified as "in a vacuum."

The space-energy hypothesis says that a classical vacuum is the absence of mass, but not the absence of energy. A point to ponder: the absolute absence of both mass and energy could mean you are outside of the Universe . . . although it may be there is no such place as outside of the Universe.

The energy content of the Universe is currently variable by 68 to 71%, the percentage range of the total energy attributed to unknown dark energy by the prevailing consensus. However, the whole question of accelerating expansion and the existence of dark energy may be moot if it is an artifact of redshift induced by a variable speed of light. Possibly the light from the early Universe does indicate it once exceeded the speed limit we know today. Perhaps the redshifts we observe and interpret as accelerating expansion are relics of a historically higher light speed. More detailed work with observational data would need to be done to determine whether the observations are consistent with the model outlined.

The Michelson-Morley experiment was looking for a media that propagates light. The concept of an additional dimension of energy was not contemplated, in the sense that has been described. We have seen that relativity negates the possibility of detecting any aether like media, but that the relationships that govern time dilation, length contraction, and relativistic mass just as readily describe a rotation into an orthogonal energy dimension.

The possibility that galactic rotation curves that seem to defy the law of gravity is related to the mass distribution within galaxies has been raised. We now know that enormous black holes are located at the centre of most galaxies, but there may be others distributed throughout space. The effect they have on the mass distribution within galaxies and the rotation curves is not completely understood.

The velocity profile of a rotating fluid could be determined in the lab or calculated mathematically using fluid dynamics software perhaps, to see if it corresponds with the relationship of stellar orbital velocities in galaxies. A comprehensive survey of the mass in galaxies may be possible someday that would account for the mass that emits light and all the mass that does not emit light or absorbs light.

Is it possible to restate Einstein's field equations in terms of space-energy instead of space-time? We have seen that Newton's first law, ($F = ma$), can be restated without reference to time as ($F = k \times \dfrac{dE}{dx}$), and that the time dimension term of Minkowski space $(ict)^2$ can be reconfigured with an energy related term $(i\lambda)^2$. Could similar modifications be made to Einstein's equations? This is a question for professional mathematicians and physicists.

Quantum mechanics remains enigmatic, although it works perfectly well in advanced technologies. The attitude of "shut up and calculate" describes the situation aptly. Is there a reality behind the probabilities, and will we ever be able to measure it? Einstein thought so, and he had an impressive track record of scientific intuition.

The behaviour of light in prisms and windows could be investigated in an optical laboratory. Measurements of light frequencies and paths through a prism could be undertaken. This might prove or disprove the intuitive proposition that light takes the shortest path through space-energy.

As outlined in the preceding chapter, there is data that can be used to evaluate the idea of variable light speed in gravity wells. The Pioneer anomaly, the observed anomalous deceleration of the Pioneer 10 and 11 spacecraft as they leave the solar system might be explained by a variation in the speed of light in a reduced energy density gravity well. Calculations based on the paradigm in this book do not match exactly the observed behaviour but may be close enough to warrant a more detailed investigation.

Other tests might be conceived by readers of this book. Or there might be a flaw in this whole hypothetical structure that relegates it to the dustbin instantly. If that is the case, *c'est la science!*

As far as how the Universe will end, we will just have to wait and c!

THE FINAL FRONTIER 20

The closing theme remains the same as the music that has played throughout this story. Perhaps by employing skeptical scrutiny to the current pronouncements of physics and cosmology and looking for the simple explanations for what we observe, we can break through the conundrums we currently face.

I imagine there will be skeptical scrutiny of everything in this book, or at least, I hope so. Some of the concepts may be considered simple restatement of what is already known. However, I believe there are sufficient new and testable ideas that should provide some intrigue.

One supposed problem with a leading-edge scientific idea called string theory, the proposal that fundamental particles are tiny vibrating strings, is that it makes no testable predictions. Several ideas that have been outlined in this story can be experimentally assessed, and certain predictions do follow from them. So, it seems appropriate to discuss some of the significant expectations for the future if the space-energy paradigm is valid, in closing.

Since the theory is based upon variable speed of light (a VSL theory) the fundamental prediction is that the speed of light in a vacuum will be found to vary, although it will not vary much over a brief period. Many years may be needed to determine whether the speed does vary over the span of the evolution of the Universe. The speed of light should be found to decline as the expansion of the Universe continues, as the density of the mass-energy it contains and conserves is spread through an ever-larger volume.

Other predictions are of a rather more negative nature. Extraordinary dark matter will not be found. Cosmic inflation caused by mysterious antigravity or dark energy will be found to be unnecessary and replaceable by extremely high light speed at the birth of the Universe. Antigravity, other than adding energy in a way that overcomes gravity, will never be discovered.

I hope and believe the laws of physics as we currently understand them will be validated and new understanding will be added to our arsenal of knowledge. The need for unknown dark matter and dark energy to explain 96% of the Universe may disappear, and we will be back on the trail of comprehensibility that Albert Einstein walked.

The four fundamental forces will be unified in that they will all be found to be caused by some form of energy gradient. Gravity resulting from density gradients in the electromagnetic background energy caused by mass, and the other three forces due to gradients in the electromagnetic field caused by charge. The mystery of gravity can potentially be explained without recourse to curved space-time. Explanations for inertia and entropy have emerged from the space-energy paradigm and perhaps they will be validated.

On the more speculative side, the "God does not play dice!" contingent will be proven correct, in that phenomena we now see as being due to waves of probability will be found to have more deterministic physical causes. The curtain will be drawn back on the mysteries of quantum mechanics. For now, quantum effects such as non-locality remain incomprehensible, but potentially lead to extremely useful technologies.

The ultimate prediction possible from cosmology concerns the fate of the entire Universe. There are many options available: an eternal cosmos of infinite duration, a heat death scenario where there are no energy gradients left to make anything happen, or a repetitive cyclical Big Bang-Big Crunch sequence.

Reasoning to support a cyclic Universe has been outlined, for consideration. The VSL hypothesis suggests the Universe will continue to expand until it stops, culminating in the ultimate black hole. If the Big Bang is a recurrent phenomenon, then presumably the contents of the Universal black hole merge, grow hot, and return to a state of pure energy of ever-increasing density, subsequently exploding in a new Big Bang, when the speed of light becomes great enough to make escape possible. Time, the ratio of distance to velocity, will tell.

If you are a professional scientist in the disciplines of cosmology or quantum physics or astronomy or engaged in any scientific pursuit and are reading this, I can only ask that you look through the lens of the unconventional paradigm offered in these pages to see if any new inspiration is created. Over the years, I have found much is to be gained by observing everything in life from a vantage point off the beaten path. If you find nothing of use in this book, hopefully it has at least been an entertaining read.

Even to disprove something is progress. Blind alleys sealed off hasten our progress along the path to the truth, and a Star Trek future. In this context, I would not be disappointed to learn that the ideas I have discussed in this book are wrong, with the accompanying reasons and scientific arguments to that effect, of course.

A general outline flowchart of the ideas in the book and the Appendix of Keywords follow and will lead you to additional resources and information on the Internet if you so desire.

Is there truly a Forth dimension? Is this science or science fiction? Yes.

THANK YOU FOR READING!

MAP OF THE THEORY

Time is a ratio of distance and motion and a means of coordinating human activities.

Space-energy can be envisioned as well as space-time.

Gravity is a result of gradients in a contoured energy background.

A fourth dimension can be envisioned.

All forces are energy gradients.

The speed of light is variable over the life of the Universe; cosmic inflation is not necessary.

Under the space-energy paradigm, Aether is alive and well—it is the energy background.

The Michelson-Morley experiment found no aether but was misinterpreted.

Dark matter is not necessary if other explanations are possible for galactic rotation curves and light deflection by the Bullet Cluster.

Dark energy is not necessary if redshift behaviour is explained by variable light speed.

Black holes do not contain singularities.

Gravitational waves are ripples in the energy background.

Mass and energy are different forms of the same thing.

The unification of the four fundamental forces may be possible.

The space energy model may explain certain aspects of quantum mechanics.

The minimal time path of Fermat's Principle may also be the minimal space-energy path.

The fate of the Universe—a cycle repeating indefinitely.

The Pioneer Anomaly: An alternative explanation.

Further work and testing of the ideas in this book.

Predictions for the Final Frontier.

ACKNOWLEDGEMENTS

This book would have been improbable without the presence of many influences and friends since my earliest days. I wish to acknowledge my parents, Frank and Eva Forth, and my brother, Ed, for making those early days a positive experience, encouraging my interests, and providing basic life support. I also thank my wife, Marla, and our three daughters—Sabrina (and her husband, Max van der Voet), Calista, and Veronica—for making family life a continuous joy, and for tolerating my odd sense of humour and writing tendencies. Special thanks to Veronica for producing most of the illustrations in the book.

I must especially thank my closest collaborator on this book, John van Leeuwen, who prepared figures, read chapters, and critiqued ideas over a period of years as the conceptual seeds sprouted and fermented. He also gave me a historical perspective on the development of science and made sure I was consistent in extending them to the new space-energy paradigm. We met in the energy industry and have been friends for more than 40 years. John and his wife Anne were also kind enough to provide a beautiful writer's cabin in the mountains of British Columbia and much great food, wine, spectacular scenery, and conversation during my stays. Thank you for your long-time friendship and encouragement!

I always had a keen interest in science and was fortunate to have as a childhood friend Terry Ingoldsby, who first showed me Mars through a small reflecting telescope and allowed me to participate in his wide-ranging interests, including astronomy, ham radio, model aircraft, steam engines, go-karts, and pizza kits.

A fortunate occurrence was my meeting up, in an alphabetic lineup in High School, with Douglas J. Fortune, who has consistently pursued scientific topics and dedicated his life to inventing a better computerized mouse trap. Even earlier, I met Allen Gommeringer, who shared my sense of humour and sharpened my wits with his original and incisive takes on all aspects of life.

My favourite teacher, now passed on, was Earl Shields, who taught High School physics at the Lethbridge Collegiate Institute in my hometown of Lethbridge, Alberta. Those classes, and subsequent engineering courses at the University of Lethbridge and the University of Alberta, were fascinating revelations into the workings of the physical world.

During my career in the petroleum industry, or energy industry as it should become, I had the privilege of working for two extremely talented bosses, Michael Danyluk and Tony van Winkoop, who excelled at their professions and as mentors. A special thanks to Mike for his ongoing encouragement and interest in the ideas as they developed.

Over the years, I have been lucky to meet up with kindred spirits on social media. They helped by asking questions and causing me to travel different paths of thought than I would have on my own. In particular, Alexandria Masiak and Kelly Lynne Bowell (Belfast) provided comments that inspired various improvements to the book.

I am grateful for the interest and comments provided by friends, including Richard van Appelen, Eldan Bosik, Les Goldstrom, and Gary Borkristl—a fellow sky watcher. There are many others out there who assisted, knowingly or not, in shaping this book, but the available space-energy precludes mentioning each of you. My gratitude and recognition go out to all my friends, family, and fellow humans who have contributed in positive ways to the realization of this retirement project and lifelong ambition.

There are dozens of authors that fed my love of science and reading over the years. Their contributions were not to this book directly, but they moulded my mind and made me want to write a book. There are too many to list completely, but the list would have to include science popularizer Carl Sagan, science fiction authors Robert A. Heinlein, Arthur C. Clarke, and Isaac Asimov, and nature writer Edward Abbey. As always, too many books and too little space or time.

Thanks to the people at FriesenPress who made the book into a physical reality.

I extend a preliminary expression of gratitude to those who will perhaps read and critique this book to determine if it has scientific value or if it is just a fictional curiosity. Regardless of how it turns out, this has been a fun and educational experience on my lifelong learning journey.

Finally, credit must be given to the wonderful online resources that are available these days, thanks to DARPA B. Internet (1983–). Younger people have no idea how lucky they are to have most of human knowledge at their fingertips, and a means to post cat videos too.

Although efforts have been made to eliminate them, any errors, omissions, incorrect ideas, or bad jokes are the responsibility of the author.

SUGGESTED READING

Carroll, Sean. *The Biggest Ideas in the Universe: Space, Time, and Motion*. New York: Dutton Penguin Random House LLC, 2022.

Davies, Paul. *The Last Three Minutes: Conjectures About the Ultimate Fate of the Universe*. New York: Basic Books, 1997.

Forth, Ron. "An Alternative Approach to Risk Analysis Recognizing the Limitations of Conventional Probabilistic Methods." Paper presented at the SPE Hydrocarbon Economics and Evaluation Symposium, Dallas, Texas (March 1997). https://doi.org/10.2118/37937-MS.

Gleick, James. *Chaos: Making a New Science*. London: Vintage, 1996.

Greene, Brian. *The Fabric of the Cosmos: Space, Time, and the Texture of Reality*. New York: Vintage Books, 2004.

Halliday, David, Robert Resnick, and Jearl Walker. The Fundamentals of Physics, sixth edition. Hoboken: John Wiley & Sons, Inc, 2001.

Isaacson, Walter. *Einstein: His Life and Universe*. New York: Simon and Schuster, 2007.

Magueijo, João. "New Varying Speed of Light Theories." *Reports on Progress in Physics* 66, no. 11 (November 2003): 2025. https://arxiv.org/pdf/astro-ph/0305457.pdf.

Sofue, Yoshiaki. "Rotation Curve Decomposition for Size-Mass Relations of Bulge, Disk, and Dark Halo in Spiral Galaxies." *Publications of the*

Astronomical Society of Japan 68, no. 1 (February 2016). https://arxiv.
org/pdf/1510.05752.pdf.

Taylor, Edwin F., and John Archibald Wheeler. *Spacetime
Physics: Introduction to Special Relativity.* San Francisco
and London: W. H. Freeman and Company,
1963. https://ia801301.us.archive.org/22/items/
SpacetimePhysicsIntroductionToSpecialRelativityTaylorWheelerPDF/
Spacetime%20Physics%20-%20Introduction%20to%20Special%20
Relativity%20[Taylor-Wheeler]PDF.pdf.

Turyshev, Slava G., Viktor T. Toth, Gary Kinsella, Siu-Chun Lee, Shing
M. Lok, and Jordan Ellis. "Support for the Thermal Origin of the
Pioneer Anomaly." *Physical Review Letters* 108, no. 24 (June 2012):
241101–6. https://arxiv.org/pdf/1204.2507.pdf.

APPENDIX OF KEYWORDS
CHOSEN FOR INTERNET SEARCHERS

Acceleration

Aether

Albert Einstein

Angular acceleration

Apsidal precession

Atomic clock

Axion

Big Bang

Black hole

Bohr radius

Brown dwarf

Calculus

Cartesian coordinates

Centripetal force

Cepheid variable star

Classical electron radius

Classical physics

Clock

Complex number

Conservation of energy

Cosmic inflation

Cosmic microwave background (CMB)

Cosmology

Curvature

Dark energy

Dark matter

Density

Differential calculus

Doppler effect

Einstein field equations

Electromagnetic spectrum

Electromagnetic wave

Electron

Energy

Entangled particles

Entropy

Equilateral triangle

Escape velocity

Fine structure constant

Fireworks

Force

Four-dimensional Euclidean sphere

Fraunhofer lines

Galactic rotation curve

Galaxy

Gradient

Gravitational potential energy

Gravitational waves

Graviton

Gravity

Gravity waves

Gravity well

Hidden variable theory

Hitchhiker's Guide to the Galaxy

Homing pigeon

Horizon problem

Hypotenuse

Hypothesis

Holy Grail

Infinity

Information

Isaac Newton

Isentropic bulk modulus

Kinetic energy

Logic

Longitudinal wave

Lorentz transformation

Mass

Max Planck

Mercury

Michelson-Morley experiment

Momentum

Moon

Nature

Neutrino

Neutron

Newton's cannonball

Newton's law of gravity

Nuclear fusion

Oblate spheroid

Orbit

Orbital eccentricity

Orbital velocity

Orthogonal

Paradigm

Photon

Physics

Potential energy

Pressure

Projection

Proof

Proton

Pythagorean theorem

Quantum physics

Quark

Quasar

Random

Reality

Redshift

Reference frame

Right triangle

Satellite

Scalar

Schwarzschild radius

Science

Second

Sensory deprivation tank

Sine wave

Singularity

Space-time

Speed of light

Speed of sound

Sphere

Spheroid

Standing wave

Star

Telescope

Temperature

Tensor

Theory

Theory of relativity

Thought experiments

Time

Tornado

Transverse wave

Twin paradox

Vacuum

Variable speed of light

Vector

Velocity

WIMPs

Wookieepedia

Work

www.ingramcontent.com/pod-product-compliance
Lightning Source LLC
Chambersburg PA
CBHW040953170526
45159CB00014B/3120